Studies in Systems, Decision and Control

Volume 310

The series "Studies in Systems, Decision and Control" (SSDC) covers both new developments and advances, as well as the state of the art, in the various areas of broadly perceived systems, decision making and control–quickly, up to date and with a high quality. The intent is to cover the theory, applications, and perspectives on the state of the art and future developments relevant to systems, decision making, control, complex processes and related areas, as embedded in the fields of engineering, computer science, physics, economics, social and life sciences, as well as the paradigms and methodologies behind them. The series contains monographs, textbooks, lecture notes and edited volumes in systems, decision making and control spanning the areas of Cyber-Physical Systems, Autonomous Systems, Sensor Networks, Control Systems, Energy Systems, Automotive Systems, Biological Systems, Vehicular Networking and Connected Vehicles, Aerospace Systems, Automation, Manufacturing, Smart Grids, Nonlinear Systems, Power Systems, Robotics, Social Systems, Economic Systems and other. Of particular value to both the contributors and the readership are the short publication timeframe and the world-wide distribution and exposure which enable both a wide and rapid dissemination of research output.

** Indexing: The books of this series are submitted to ISI, SCOPUS, DBLP, Ulrichs, MathSciNet, Current Mathematical Publications, Mathematical Reviews, Zentralblatt Math: MetaPress and Springerlink.

More information about this series at http://www.springer.com/series/13304

Jun Fu · Ruicheng Ma

Stabilization and H_∞ Control of Switched Dynamic Systems

Springer

Jun Fu
The State Key Laboratory of Synthetical
Automation for Process Industries
Northeastern University
Shenyang, China

Ruicheng Ma
School of Mathematics
Liaoning University
Shenyang, China

ISSN 2198-4182 ISSN 2198-4190 (electronic)
Studies in Systems, Decision and Control
ISBN 978-3-030-54199-6 ISBN 978-3-030-54197-2 (eBook)
https://doi.org/10.1007/978-3-030-54197-2

This Springer imprint is published by the registered company Springer Nature Switzerland AG
The registered company address is: Gewerbestrasse 11, 6330 Cham, Switzerland

Acknowledgements

There are numerous individuals without whose constructive comments, useful suggestions and wealth of ideas this monograph could not have been completed. Special thanks go to Prof. Tianyou Chai at Northeastern University, Prof. Wei Lin at Case Western Reserve University, Dr. Jun Liu at the University of Sheffield, Prof. Frank Lewis at the Texas University at Arlington, Prof. Zhentao Hu at Henan University, Prof. Shengzhi Zhao at Liaoning University, and Associate Professor Yan Liu at Northeastern University. Thanks also go to our students, Jinghan Li, Shuang An, for their commentary. The authors are especially grateful to their families for their encouragement and never-ending support when it was most required. Finally, we would like to thank the editors at Springer for their professional and efficient handling of this project.

The writing of this book was supported in part by National Natural Science Foundation of China (61825301, 61673198, 62073157), Provincial Natural Science Foundation of Liaoning Province (20180550473), Scientific Research Fund of Educational Department of Liaoning Province (LZD201901) and Liaoning Revitalization Talents Program (XLYC1807012).

Shenyang, China
June 2020

Jun Fu
Ruicheng Ma

Contents

Chapter 1
Introduction

1.1 Switched Systems

Switched systems constitute a special class of hybrid systems which contain both continuous dynamics and discrete dynamics. Switched systems arise in engineering practice where several dynamical system models are required to model an engineering system due to the various jumping parameters and changing environmental factors [1]. In addition, switching control between different controllers has proved powerful technique for solving problems in, for example, the stabilization of the cart-pendulum system, when no single controller is effective [2]. Therefore, it has very important scientific significance and application value for the research of switched systems. A number of researchers have considered switched systems because of their importance from both theoretical and practical points of view (see, e.g., [3–7] and the references therein).

The dynamic behavior of switched system is extremely complex, mainly due to the coexistence of continuous dynamic and discrete switching signals and their interaction. Even though all subsystem of switched system are linear, it should be classified as a nonlinear system in nature, because the introduction of discrete switching signals is beyond the scope of linear systems, and its dynamic behavior is even more general. For example, switching among different system modes make a switched systems display the phenomena of chaos, Zeno, and multiple limit cycles, etc. For the stability of a switched system, switching among stable subsystems may lead to unstability of the whole switched system; conversely, switching among unstable subsystems may lead to stability of the whole switched system. Therefore, the switching signal plays a very important role in switched systems and can yield the whole switched system to have properties that all the subsystem do not have.

Let us take the following interesting examples.

J. Fu and R. Ma, *Stabilization and H_∞ Control of Switched Dynamic Systems*, Studies in Systems, Decision and Control 310,
https://doi.org/10.1007/978-3-030-54197-2_1

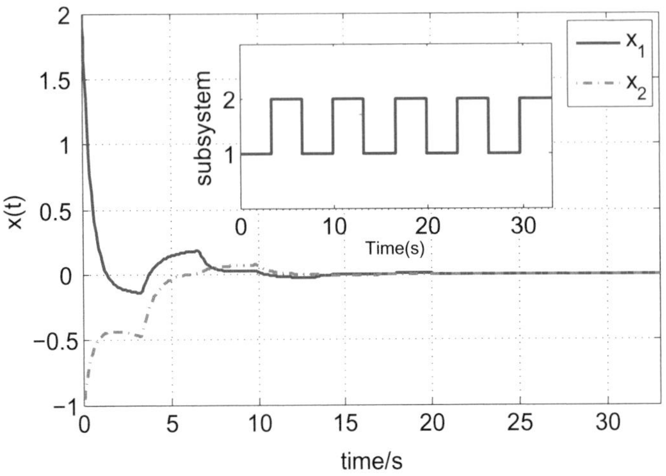

Fig. 1.1 State responses and switching signal

Example 1.1 [8] Consider the switched linear system

$$\dot{x} = A_\sigma x \tag{1.1}$$

composed of two subsystems with the following system matrices
with $A_1 = \begin{bmatrix} -1.9 & 0.6 \\ 0.6 & -0.1 \end{bmatrix}$, $A_2 = \begin{bmatrix} 0.1 & -0.9 \\ 0.1 & -1.4 \end{bmatrix}$.

It is clear that both subsystems are unstable, since the eigenvalues of A_1 and A_2 are $\{-2.0817, 0.0817\}$ and $\{0.0374, -1.3374\}$, respectively.

It has been shown in [9] that, the switched system is stable under the dwell time switching with the range [0.6, 3.1]. The state responses of the switched system under a random switching signal are depicted in Fig. 1.1 with the initial state $(2, -1)^T$.

Example 1.2 [10] Consider the continuous stirred tank reactor (CSTR) with two modes feed stream (see Fig. 1.2), which is molded as a switched nonlinear system:

$$\begin{aligned} \dot{C}_A &= \frac{q_\sigma}{V}\left(C_{Af_\sigma} - C_A\right) - a_0 e^{-E/(RT)} C_A, \\ \dot{T} &= \frac{q_\sigma}{V}\left(T_{f_\sigma} - T\right) - a_1 e^{-E/(RT)} C_A + a_2\left(T_c - T\right), \end{aligned} \tag{1.2}$$

where C_A is the reactant A concentration, T is the reactor temperature, $\sigma : [0, +\infty) \to M = \{1, 2\}$ is the switching signal, q_i is the feed flow rate, C_{Af_i} is the concentration, T_{f_i} is the temperature, V is the volume of the reactor, E is the activation energy, T_c is the coolant temperature, R is the gas constant, a_0, a_1, a_2 are constant coefficients.

Stability is an important property of switched systems [11–15]. However, stability analysis is quite difficult due to the coexistence of continuous dynamic and discrete

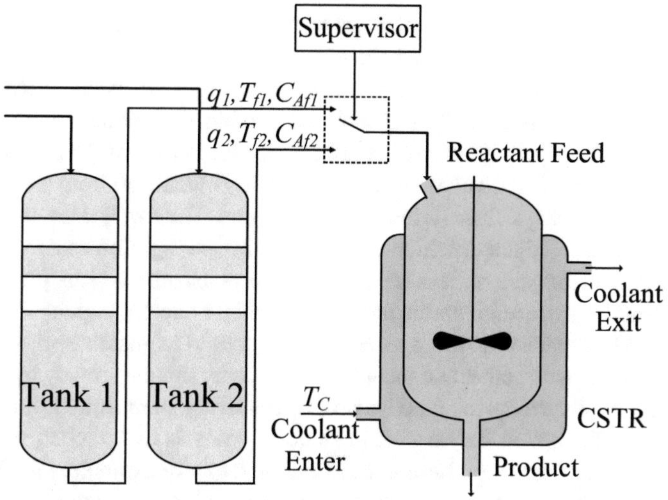

Fig. 1.2 Schematic diagram of the process

switching signals and their interaction. Reference [1] surveyed the developments in three basic problems regarding stability and design of switched systems:

- stability for arbitrary switching sequences,
- stability for certain useful classes of switching sequences,
- construction of stabilizing switching sequences.

Stability under arbitrary switchings is a desirable property of switched systems due to its practical importance. Some efforts have been made finding conditions that guarantee asymptotic stability of a switched system for arbitrary switching signals. The existence of a common Lyapunov function for all subsystems was shown to be a necessary and sufficient condition for a switched system to be asymptotically stable under arbitrary switchings [4]. A number of conditions have been given towards the existence of a common Lyapunov function, mainly for switched linear systems, e.g., see [16, 17]. Even though some progress has been made in, such as, [18, 19], finding a common Lyapunov function for a family of subsystems, especially for switched nonlinear systems, is still an open problem unless they are in some particular form. For instance, [20] presented constructions of a common Lyapunov function for a finite family of pairwise commuting globally asymptotically stable nonlinear systems. In the recent years, switched nonlinear systems in lower triangular form has also drawn considerable attention to study the global stabilization under arbitrary switchings [21, 22] and some designed switching signal [23, 24].

On the other hand, most switched systems in practice do not possess a common Lyapunov function, yet they still may be asymptotically stable under some properly chosen switching law. Therefore, if a switched system is not asymptotically stable for arbitrary switching, one should identify those switching signals for which it is asymptotically stable. The multiple Lyapunov function technique proposed by [25] has proven to be a powerful and effective tool for finding such a switching law and many results in this direction have been available [3, 7, 26].

It is a fairly well known fact that when all subsystems are asymptotically stable, the switched system is globally exponentially stable if their active dwell time is enough long to dissipate the transient effect after each switching [27]. In fact, the required lower bound on dwell time can be explicitly calculated from the parameters of the individual subsystems. Reference [28] extends the dwell time concept to an average dwell time concept, which means that the average time interval between consecutive switchings is no less than a specified constant, and it is proved that, if such a constant is sufficiently large, then the switched system is exponentially stable. Reference [29] extended the average dwell time method to guarantee the stability of switched systems with stable and unstable subsystems simultaneously by restricting these unstable subsystems to be activated in a relatively short time. It turns out that the idea of dwell time is very useful in stability analysis of switched systems and many results in this direction have been available (see, for example, [30–37]).

When all the individual subsystems are unstable, it is possible to find a switching signal that renders the switched system asymptotically stable. Several methods have been proposed to construct this switching signal [4, 38]. A typical switching law is given by the state-dependent switching strategies such as the convex combination method [4], the min-projection strategy [39], largest region function strategy [38], etc. Another switching law is given by the time-dependent switching strategies particularly concerned with dwell time. Based on the dwell time switching, [8] exploited the stabilization property of switching behaviors to compensate the state divergence made by unstable modes, and then proposed a sufficient condition ensuring the asymptotic stability of switched systems with all modes unstable. Then, based on the dwell time switching, [40, 41] investigated the stabilization of continuous-time switched positive systems by finding stabilizing switching laws.

So far, the basic framework of the theory of switched systems has been established. Some research results on switched systems have been applied to multi-agent systems [13], chemical process control systems [10], etc. However, it should be pointed out that the theory of switched systems needs to be further improved. Thus, attention is shifted to switched nonlinear systems with special structures as system structures play an important role in the stabilization of switched nonlinear systems [21, 22, 42, 43].

1.2 Finite-Time Stabilization

In recent years, the finite-time control is one of interesting topics in control theory. In many applications, a dynamical system is desired to possess the property that trajectories converge to a Lyapunov stable equilibrium in finite-time rather than merely asymptotical. Therefore, the objective of finite-time control is to design a control law making the system state converge to the origin in finite time. The concept of finite-time stability arises naturally in time optimal control. A classical example is double integrator with bang–bang time-optimal feedback control. The problem of finite-time stabilization is one of the most important problems of finite-time control and has attracted increasing attention in recent years in, for example, ([44–46] and the references cited therein) because the finite-time stabilized systems usually practically demonstrate some desired features such as faster convergence rates, higher accuracies, and better disturbance rejection properties [44]. Because of these significant advantages, the finite-time control has been widely used in engineering, for example, in missile systems, communications network systems, robot manipulation and so on.

The authors of [47] proposed a Lyapunov stability theorem for finite-time stability analysis of continuous autonomous systems, which also provided a basic tool for synthesis of nonlinear control systems and then various methods have been proposed for a number of classes of systems (see [48–60] and references therein). Several fundamental design approaches, such as backstepping [61], adding a power integrator technique [44, 62], homogeneous domination approach [63, 64] and dynamic gain control design method [45], are deployed to solve the finite-time stabilization problem of nonlinear systems with special structures such as strict-feedback form, p-normal form and their variants [65–68] and their applications such as the underactuated unstable mechanical system introduced in [69]. Reference [44] obtained the finite-time stabilization result by combining the finite-time stability results with adding one power integrator technique. Reference [61] considered the problem of global finite-time stabilization for a class of triangular nonlinear systems based on backstepping and dynamic exponent scaling. References [44, 64] proposed the explicit design scheme of finite-time stable controller for two classes of nonlinear systems. [62] presented the adaptive finite-time control methodology for a class of uncertain nonlinear control systems. Reference A global nonsmooth stabilization scheme is presented for a class of nonlinear cascaded systems with uncontrollable linearization in [70]. Naturally, a fundamental and unsolved problem is: *Can the finite-time control design for non-switched nonlinear systems be extended to switched nonlinear systems?*

The finite-time control problems for switched nonlinear systems have attracted increasing attention in the recent years because the finite time control usually demonstrate some desired features. With the help of homogeneous techniques, [56] investigated finite-time stability and robust stability of a class of uncertain switched nonlinear systems. Reference [49] studied the finite-time stability of a class of hybrid dynamical systems with both switching and impulsive effect. Finite-time input-to-state stability problem for switched nonlinear systems was presented in [71]. Although these results provides some methods for studying the finite-time control for switched nonlinear systems, to the authors' best knowledge, few works has been done for the finite-time stabilization problem for switched nonlinear systems in strict-feedback form based on constructive design method. Therefore, Chap. 2 aims to address this problem.

In the past few years, asymptotic stabilization of switched nonlinear systems in lower triangular form has received much attention and a few important results have also appeared in, for example, [21, 22, 72–74]. One feature of the studied switched systems in the mentioned references above is that the powers of the chained integrators are restricted to the same positive odd integer for subsystems during the whole operation time. The authors of [75] studied the more general switched systems in p-normal form, where the powers of the subsystems are allowed to be different and to be positive even integers, and obtained a sufficient condition under which a globally asymptotic stabilizer is designed. The authors of [76] investigated the global exponential stabilization under arbitrary switchings for a class of switched nonlinear systems in power integrator triangular form, whose subsystems have chained integrators with the powers of different positive odd numbers, and proposed a systematic approach of constructing the global exponential stabilizer. However, the powers of integrators of the switched systems considered in [21, 22, 72, 73, 75, 76] can be only *positive integers*, and thus these methods are not applicable to switched systems where powers of the chained integrators are *not integers*. Questions naturally arise: is it possible to finite-time or asymptotically stabilize smooth switched nonlinear systems whose powers of integrators are not necessarily integers and also different for different periods of the whole operation time? Furthermore, what if the switched nonlinear systems contain non-smooth nonlinearities? If possible, under what conditions can we design such controllers and how? To our best knowledge, in the literature there have not been results which provide answers to these questions. Since the powers can be less than 1, the switched nonlinear system does not have to have first approximation at the origin, which makes the standard backstepping techniques not applicable. In addition, the hybrid features of switched systems make stabilization more difficult than non-switched systems, which imposes another challenge. Therefore, Chap. 3 aims to address this problem.

The synthesis for switched systems with special structure has begun to attract attention. For example, the stabilization for high-order switched systems has been studied in [73] by adding a power integrator technique, and for the switched nonlinear systems in non-triangular form but in special structure, [77] investigated the stabilization of switched systems by designing an appropriate switching law and individual subsystem controllers. However, the powers of integrators of switched systems studied in [73, 77] can be only positive integers, and thus these methods can not deal with the switched systems where powers of the chained integrators are not integers. *Is it possible to global finite-time stabilize switched systems whose powers of integrators are not necessarily integers?* On the other hand, it is very interesting to investigate the finite-time stabilization for switched systems, the structure of whose subsystems is more general, which is no longer the one in [21, 22, 72, 73, 75] but contain them as a special case. If possible, under what conditions can we design such a controller and how? To our best knowledge, in the literature there have not been results which provide answers to these questions. Therefore, Chap. 4 aims to address this problem.

On the other hand, parametric uncertainties exist in many practical systems, for which adaptive control is one of the effective ways [78, 79]. For the case of unknown parameters with *known* bounds, adaptive state-feedback controllers are developed for nonlinear systems in [80] and [81], for convex/concave parametrization and general nonlinear parametrization, respectively. For the case of unknown parameters with *unknown* bounds, several fundamental design approaches, such as backstepping, and adding a power integrator technique, are developed to solve the adaptive control problem of nonlinear systems with linear parametrization [82] and with nonlinear parametrization [68, 83], respectively. For the latter, when parameters change rapidly or abruptly, stability and convergence properties will be affected or destroyed [84] and the classical adaptive control fails in some setting mainly due to the fact that conventional adaptive control techniques provide a systematic approach for automatic adjustment of controllers in real continuous time. To cope with this case, adaptive control employing switching techniques [74, 85–87] are proposed to design adaptive stabilizers for nonlinearly parameterized systems in parametric-strict-feedback form. The significant contributions of the mentioned methods above focus on asymptotic stabilization. Naturally, a fundamental and unsolved problem is: *whether it is possible to apply the logic-based switching technique in* [74, 86, 87] *to achieve adaptive finite-time stabilization?*

In recent years, adaptive finite-time control of nonlinear systems has received much attention and a few important results have also appeared in [62, 88–90]. Reference [62] for the first time presents a fundamental result on adaptive finite-time control of p-normal form systems with positive odd integer powers. The authors of [88] and [89] studied adaptive finite-time control problems for p-normal form systems with the powers of *not less than* 1 positive odd rational numbers, and with the powers of *not bigger than* 1 positive odd rational numbers, respectively. The authors of [90] investigated semi-globally uniformly adaptive finite-time boundedness for a class of switched nonlinear systems in strict-feedback form via backstepping. However, these methods in [62, 88–90] are not applicable to p-normal form systems where powers are allowed to be *any* positive odd rational numbers. Note that the unknown parameters in [89] do not appear in the system nonlinear functions. Note also that [62, 88] require that the unknown parameters are constant. Therefore, it is interesting to investigate whether it is possible to apply the logic-based switching technique in [74, 86, 87] to achieve adaptive finite-time stabilization of a class of p-normal form nonlinear systems, whose powers of the chained integrators are allowed to be *any* positive odd rational numbers, and whose parametric uncertainties entering the state equations nonlinearly can be fast time-varying or jumping at unknown time instants, and the bounds of which are not required to be known a priori, if yes, under what conditions it can be achieved. Therefore, Chap. 5 aims to address this problem.

1.3 H_∞ Control

In real engineering system, a control system is always subject to unknown inputs in the form of internal uncertainties and exogenous loads that cannot be measured or inconvenient to measure. If these unknown inputs are not properly treated, poor performance of the control system may be resulted. To attenuate the negative impact from unknown inputs, various methods can be used.

The H_∞ performance of a control system with unknown inputs has been known to be a useful measure for stabilization and L_2-gain performance. The objective of H_∞ control is to design a control law for a system such that the closed-loop system is internally stable, meanwhile guaranteeing L_2-gain from external disturbances to its controlled output less than or equal to a certain prespecified constant. H_∞ control has been extensively studied in the literature (see, for example, [91, 92], and references therein), which can cope with many robustness problems such as sensitivity minimization [93] and stabilization of uncertain systems [94, 95]. To date, much attention has been paid to investigating the H_∞ control of switched systems since many practical applications can be formulated as H_∞ control of switched systems such as networked control systems [96] and target tracking systems [97].

For H_∞ control of switched systems, there are mainly three main methods: arbitrary switchings, state-dependent switchings, and dwell-time-based methods [7, 32, 73, 98–101]. Xu and Teo in [101] proposed a switched Lyapunov function for a class of switched impulsive systems to achieve H_∞ property under arbitrary switchings. The authors of [7, 73, 98–100] developed state-dependent switching strategies for H_∞ control problems of those switched systems with their own spectacular features. For the dwell-time-based method, the references [30–33, 35–37, 98, 102] exploited the feature of the dwell time technique: a dwell time of active subsystem can subside possible large state transients. References [98, 102] both investigated H_∞ control problems for switched linear systems in continuous time and discrete time, respectively. However, both results are under the assumption that all subsystems are either stable or stabilizable. References [31–33, 35–37] require that all subsystems are stable or at least one is stable, and only guarantee a weaker weighted (by a decreasing exponential function) disturbance attenuation level than a standard L_2-gain property. When the power of the decreasing exponential function (e.g., λ in [36]) approaches zero, the weighted disturbance attenuation level goes to the attenuation level of standard L_2-gain property. *However, the power of the exponential function (i.e., the above λ) approaching zero results that dwell time of switched systems goes to infinity.* This fact requires that there must be at least one *stable* subsystem of the switched system to achieve the internal stabilization. Although the work by Geromel and Colaneri [30] ensures the standard L_2-gain performance for a class of switched linear systems, the authors pose a conservative assumption that all subsystems of the switched system are stable. It is known that stabilization of switched continuous-time systems with all modes unstable has been achieved by the dwell time switching technique in [8, 103, 104]. Thus, questions naturally arise: is it possible, in the framework of the dwell time technique, to achieve H_∞ control of unforced or forced switched systems with the standard L_2-gain performance *without posing any internal stability requirements on subsystems of the switched systems*? If possible, under what conditions can we come up with a switching control to achieve this goal and how? To our best knowledge, in the literature there have not been results which provide answers to these questions. Therefore, Chap. 6 aims to address this problem.

1.4 Switched Observers Design

The problem of state observation for systems with unknown inputs has been widely studied during the last two decades, which is another effective method to deal with the unknown inputs. As full or partial information of state is unavailable for feedback in abundant practical applications [84, 105, 106], state estimation is an important issue for such applications. It is known that the unknown input observers have been widely used in the framework of fault detection and isolation.

In the literature there can be found only a few papers related to observer design for switched systems with unknown inputs. For the state estimation problem of switched systems, the most relevant papers are [107–116] which studied state estimation of switched systems under different switching strategies: arbitrary switchings [107], observer-driven switching [109], and periodic switching methods [110], and dwell-time-based methods [112]. However, these switched systems do not contain unknown inputs. Since the unknown inputs not only impact the switched systems directly, but also enter the error dynamics from observer design, the methods in [107–112, 117] are not applicable to switched systems containing unknown inputs.

On the other hand, since the disturbances or partial inputs can be encountered [113–115], the state observation for switched systems with unknown inputs has attracted increasing attention in recent years. To particularly focus on observers design for the unknown inputs switched systems, some methods have been proposed. [113–116] decoupled the unknown inputs from the dynamics of unavailable state component for observer designs *under the assumption that all or part of subsystems are strongly detectable*. This assumption indeed compromises to the obstacle that if none of subsystems is strongly detectable, then it is impossible to design state observers. Thus, questions naturally arise: is it possible, in the framework of the dwell time technique, to achieve the observers design of switched linear systems with unknown inputs *without posing any strong detectability requirements on subsystems of switched systems*? If possible, under what conditions can we achieve this goal and how? To our best knowledge, in the literature there have not been results which provide answers to these questions. Therefore, Chap. 7 aims to address this problem.

1.5 Basic Definitions and Lemmas

Several definitions are introduced that will be used throughout the book.

Now, we first review some terminologies and theorems about finite-time stability by considering the autonomous system:

$$\dot{x} = f(x); \quad f(0) = 0, \quad x \in R^n. \tag{1.3}$$

Definition 1.1 ([118]) Consider the nonlinear system (1.3), where $f : D \to R^n$ is non-Lipschitz continuous on open neighborhood D of the origin $x = 0$ in R^n. The equilibrium $x = 0$ of (1.3) is finite-time convergent if there are an open neighborhood U of the origin and a function $T_x : U \backslash \{0\} \to (0, \infty)$, such that every solution trajectory $x(t, x_0)$ of (1.3) starting from the initial point $x_0 \in U \backslash \{0\}$ is well-defined and unique in forward time for $t \in [0, T_x(x_0)]$, and $x(t, x_0) \to 0$ as $t \to T_x(x_0)$. Here, $T_x(x_0)$ is called the settling time (of the initial state x_0). The equilibrium $x = 0$ of (1.3) is finite-time stable if it is Lyapunov stable and finite-time convergent. If $U = D = R^n$, the origin is a globally finite-time stable equilibrium.

To achieve our control objective, we introduce the following lemmas.

Lemma 1.2 ([44]) *Consider the system (1.3). Suppose these are C^1 function $V(x)$ defined on a neighborhood $\mathrm{D} \subset R^n$ of the origin, and real numbers $k > 0$ and $\alpha \in (0, 1)$, such that (1) $V(x)$ is positive definite on D; (2) $\dot{V} + kV^{\alpha} \leq 0, \forall x \in D$. Then, the origin of the nonlinear system (1.3) is locally finite-time stable. The settling time, depending on the initial state $x(0) = x_0$, satisfies $T_x(x_0) \leq \frac{V(x_0)^{1-\alpha}}{k(1-\alpha)}$ for all x_0 in some open neighborhood of the origin. If $\mathrm{D} = R^n$ and $V(x)$ is also radially unbounded, the origin of the nonlinear system (1.3) is globally finite-time stable.*

Lemma 1.3 ([119]) *For any positive real numbers a, b and any real valued function $\omega(x, y) > 0$, then $|x|^a|y|^b \leq \frac{a}{a+b}\omega(x, y)|x|^{a+b} + \frac{b}{a+b}\omega(x, y)^{-a/b}|y|^{a+b}$.*

Lemma 1.4 ([44]) *For any real numbers x_i, $i = 1, \cdots, n$ and $0 \leq a \leq 1$, then $(|x_1| + \cdots + |x_n|)^a \leq |x_1|^a + \cdots + |x_n|^a$. When $a = \frac{c}{d} \leq 1$, where $c > 0$ and $d > 0$ are odd integers, then $|x^a - y^a| \leq 2^{1-a}|x - y|^a$.*

To study finite-time stabilization, we will recall some essential concepts and lemmas that will serve as the basis for the development of our finite-time stabilizers, and then give the system description.

Consider the nonlinear time-varying system

$$\dot{x}(t) = f(t, x(t)); x(t_0) = x_0, t \in I_{x_0,t_0}, \tag{1.4}$$

where $x(t) \in \mathcal{D}$, $t \in I_{x_0,t_0}$ is the system state vector, $I_{x_0,t_0} \triangleq [t_0, \tau_{x_0,t_0})$, $t_0 < \tau_{x_0,t_0} \leq \infty$, is the maximal interval of existence of a solution $x(t)$ of (1.4), $\mathcal{D} \subseteq R^n$ is an open set with $0 \in \mathcal{D}$, $f : [0, \infty) \times \mathcal{D} \to R^n$ is such that $f(\cdot, \cdot)$ is jointly continuous in t and x, and for every $t \in [0, \infty)$, $f(t, 0) = 0$. $s(t, t_0, x_0)$ is a solution of (1.4) starting from x_0 at t_0 satisfying the consistency property and the semi-group property described in [120]. We assume that (1.4) possesses unique solutions in forward time for all initial conditions except possibly the origin.

Definition 1.5 (*Finite-Time Lyapunov Stability* [120]) The zero solution $x = 0$ to (1.4) is finite-time stable if there is an open neighborhood $\mathcal{N} \subseteq \mathcal{D}$ of the origin and a function $T : [0, \infty) \times \mathcal{N} \setminus \{0\} \to [0, \infty)$, called the settling-time function, such that the following statements hold:
(i) Finite-time convergence. For every $t_0 \in [0, \infty)$ and $x_0 \in \mathcal{N} \setminus \{0\}$, $s(t, t_0, x_0)$ is defined on $[t_0, T(t_0, x_0))$, $s(t, t_0, x_0) \in \mathcal{N} \setminus \{0\}$ for all $t \in [t_0, T(t_0, x_0))$, and

$\lim_{t \to T(t_0, x_0)} s(t, t_0, x_0) = 0$.
(ii) Lyapunov stability. For every $\varepsilon > 0$ and $t_0 \in [0, \infty)$, there exists $\delta = \delta(\varepsilon, t_0) > 0$ such that $\mathcal{B}_\delta(0) \subseteq \mathcal{N}$ and for every $x_0 \in \mathcal{B}_\delta(0) \setminus \{0\}$, $s(t, t_0, x_0) \in \mathcal{B}_\varepsilon(0)$ for all $t \in [t_0, T(t_0, x_0))$.
The zero solution $x(t) = 0$ to (1.4) is globally finite-time stable if it is finite-time stable with $\mathcal{N} = \mathcal{D} = R^n$.

Lemma 1.6 (Finite-Time Lyapunov Stability Theorem [120]) *For system (1.4), the following statements hold:*
(i) If there exist a continuously differentiable function $V : [0, \infty) \times \mathcal{D} \to R$, a class κ function $\alpha(\cdot)$, and a function $\bar{k} : [0, \infty) \times R_+$ such that $\bar{k}(t) > 0$ for almost all $t \in [0, \infty)$, a real number $\lambda \in (0, 1)$, and an open neighborhood $\mathcal{M} \subseteq \mathcal{D}$ of the origin such that

$$V(t, 0) = 0, t \in [0, \infty), \tag{1.5}$$

$$\alpha(\|x\|) \leq V(t, x), t \in [0, \infty), x \in \mathcal{M}, \tag{1.6}$$

$$\dot{V}(t, x) \leq -\bar{k}(t) V(t, x)^\lambda, t \in [0, \infty), x \in \mathcal{M}, \tag{1.7}$$

then the zero solution $x(t) = 0$ to (1.4) is finite-time stable.
(ii) If $\mathcal{N} = \mathcal{D} = R^n$ and there exist a continuously differentiable function $V : [0, \infty) \times \mathcal{D} \to R$, a class κ_∞ function $\alpha(\cdot)$, a function $k : [0, \infty) \to R_+$ such that $k(t) > 0$ for almost all $t \in [0, \infty)$, and an open neighborhood $\mathcal{M} \subseteq \mathcal{D}$ of the origin such that (1.5)–(1.7) hold, then $x(t) = 0$ to (1.4) is globally finite-time stable.

Furthermore, we will frequently need the following inequalities adopted from [44].

- For $x, y \in R$ and $1 \leq q \in Q_{odd}$, we have

$$| x + y | \leq 2^{1-1/q} \mid x^q + y^q \mid^{1/q} . \tag{1.8}$$

- For any real numbers $a > 0$, $b > 0$ and any real valued function $\omega(x, y) > 0$, the following inequality holds:

$$|x|^a |y|^b \leq \frac{a}{a+b} \omega(x, y) |x|^{a+b} + \frac{b}{a+b} \omega(x, y)^{-\frac{a}{b}} |y|^{a+b}. \tag{1.9}$$

- For $a, b, c \in R$, if $0 < a \leq b \leq c$, it is true that

$$|x|^b \leq |x|^a + |x|^c = |x|^a (1 + |x|^{c-a}), x \in R. \tag{1.10}$$

To study finite-time stabilization, we introduce some basic concepts and lemmas related to the notion of finite-time stability and the corresponding Lyapunov stability theory. We then formulate the problem of global finite-time stabilization of switched system (1.6).

Lemma 1.7 (Finite-Time Lyapunov Stability Theorem [47]) *For system (1.4), suppose there exist a C^1 positive definite function $V : U \rightarrow R$ defined on an open neighborhood U of the origin such that $\dot{V} + cV^{\alpha} \leq 0$ on U for some $c > 0$ and $\alpha \in (0, 1)$. Then the origin of system (1.4) is a finite-time stable equilibrium and the settling time T satisfies $T(x_0) \leq \frac{V(x_0)^{1-\alpha}}{c(1-\alpha)}$ for the initial state x_0. If $U = R^n$ and $V(x)$ is radially unbounded, the origin of system (1.4) is globally finite-time stable equilibrium.*

Furthermore, we will frequently need the following inequalities adopted from [44].

• For $x, y \in R$ and $1 \leq q \in Q_{odd}$, we have

$$| x + y | \leq 2^{1-1/q} \mid x^q + y^q \mid^{1/q} . \tag{1.11}$$

• For any real numbers $a > 0$, $b > 0$ and any real valued function $\omega(x, y) > 0$, the following inequality holds:

$$|x|^a |y|^b \leq \frac{a}{a+b} \omega |x|^{a+b} + \frac{b}{a+b} \omega^{-\frac{a}{b}} |y|^{a+b}. \tag{1.12}$$

• For $a, b, c \in R$, if $0 < a \leq b \leq c$, it is true that

$$|x|^b \leq |x|^a + |x|^c = |x|^a (1 + |x|^{c-a}), x \in R. \tag{1.13}$$

1.6 Organization of the Book

This book study the finite-time stabilization and H_∞ control of switched systems. Structure of the book is summarized as follows.

This chapter has introduced the system description and some background knowledge, and also addressed the motivations of the book.

In Chap. 2, the global finite-time stabilization problem for a class of switched nonlinear systems under arbitrary switchings is investigated. All subsystems of the studied switched system under consideration are in lower triangular form. Based on the adding one power integrator technique, both a class of non-Lipschitz continuous state feedback controller and a common Lyapunov function are simultaneously constructed such that the closed-loop switched system is global finite-time stability under arbitrary switchings. In the controller design process, a common coordinate transformation of all subsystems is exploited to avoid using individual coordinate transformation for each subsystem. Finally, Two examples are given to show the effectiveness of the proposed method.

In Chap. 3, the global finite-time stabilization is studied for a class of switched strict-feedback nonlinear systems, whose subsystems have chained integrators with the powers of positive odd rational numbers (i.e., numerators and denominators of the powers are all positive odd integers but not necessarily relatively prime). All the powers in each equation of subsystems of the switched systems can be different. Based on the technique of adding a power integrator, the global finite-time stabilizers of individual subsystems are first systematically constructed to guarantee global finite-time stability of the closed-loop smooth switched system under arbitrary switchings, and then a co-design of stabilizers and a state-dependent switching law is proposed to achieve global finite-time stabilization of the closed-loop non-smooth switched systems. In the controller design, a common coordinate transformation of all subsystems is exploited to avoid using individual coordinate transformations for individual subsystems. We also give some sufficient conditions that enable our design by characterizing the powers of the chained integrators of the considered switched systems. Numerical examples are provided to demonstrate the effectiveness of the proposed results.

In Chap. 4, the global finite-time stabilization is investigated for a class of switched nonlinear systems in non-triangular form, whose subsystems have chained integrators with the powers of positive odd rational numbers (i.e., numerators and denominators of the powers are all positive odd integers). All subsystems are not assumed to be stabilizable. Based on the technique of adding a power integrator and the multiple Lyapunov functions method, both the global finite-time stabilizers of individual subsystems and a switching law are systematically constructed to guarantee global finite-time stabilization of the closed-loop switched nonlinear system. A numerical example is provided to illustrate the effectiveness of the proposed method.

In Chap. 5, the global adaptive finite-time stabilization is investigated by logic-based switching control for a class of uncertain nonlinear systems with the powers of positive odd rational numbers. Parametric uncertainties entering the state equations nonlinearly can be fast time-varying or jumping at unknown time instants, and the control coefficient appearing in the control channel can be unknown. The bounds of the parametric uncertainties and the unknown control coefficient are not required to know a priori. Our proposed controller is a switching-type one, in which a nonlinear controller with two parameters to be tuned is first designed by adding a power integrator, and then a switching mechanism is proposed to tune the parameters online to finite-time stabilize the system. An example is provided to demonstrate the effectiveness of the proposed result.

In Chap. 6, the standard H_∞ control of switched systems is studied via dwell time switchings without posing any internal stability requirements on subsystems of the switched systems. First, a sufficient condition is formed by specifying lower and upper bounds of the dwell time, constraining upper bound of derivative of Lyapunov function of the active subsystem, and forcing the Lyapunov function values of the overall switched system to decrease at switching times to achieve standard H_∞ control of unforced switched linear systems. Then, in the same framework of the dwell time, sufficient conditions are given for that of the corresponding forced switched linear systems by further designing state feedback controllers. Finally, numerical examples are provided to demonstrate the effectiveness of the proposed results.

In Chap. 7, the state observers design of a class of unknown inputs switched linear systems is investigated via mode-dependent dwell time switchings. The distinguishing feature of the proposed method is that strong detectability condition of subsystems of the switched systems is unnecessarily required. Firstly, a time-varying coordinate transformation is introduced to design a suitable reduced-order observer for each subsystem. Then, computable sufficient conditions on the synthesis of the observers are proposed in the framework of a mode-dependent dwell time technique. Since the observer of individual subsystem cannot be designed due to unavailability of strong detectability condition of the subsystem, the state of the switched system is estimated under the condition of confining the dwell time by a pair of upper and lower bounds, restricting the growth of Lyapunov function of the active subsystem, and forcing "energy" of the overall switched system to decrease at *switching instants*. Next, we apply our method to the stabilization of switched singular linear systems. Finally, examples are presented to demonstrate the effectiveness of the proposed methods.

In Chap. 8, we conclude the monograph by briefly summarizing the main theoretical findings presented in our book, and proposing unsolved problems for further investigations.

References

1. Liberzon, D., Morse, A.S.: Basic problems in stability and design of switched systems. IEEE Control Syst. Mag. **19**(5), 59–70 (1999)
2. Zhao, J., Spong, M.W.: Hybrid control for global stabilization of the cart-pendulum system. Automatica **37**(12), 1941–1951 (2001)
3. Branicky, M.S.: Multiple lyapunov functions and other analysis tools for switched and hybrid systems. IEEE Trans. Autom. Control **43**(4), 475–482 (1998)
4. Liberzon, D.: Switching in Systems and Control. Birkhauser, Boston (2003)
5. Lin, H., Antsaklis, P.J.: Stability and stabilizability of switched linear systems: a survey of recent results. IEEE Trans. Autom. Control **54**(2), 308–322 (2009)
6. Sun, Z., Ge, S.S.: Switched Linear Systems: Control and Design. Springer, London (2005)
7. Zhao, J., Hill, D.J.: On stability, L_2-gain and H_∞ control for switched systems. Automatica **44**(5), 1220–1232 (2008)
8. Xiang, W., Xiao, J.: Stabilization of switched continuous-time systems with all modes unstable via dwell time switching. Automatica **50**(3), 940–945 (2014)
9. Fu, J., Ma, R., Chai, T., Hu, Z.: Dwell-time-based standard H_∞ control of switched systems without requiring internal stability of subsystems. IEEE Trans. Autom. Control **64**(7), 3019–3025 (2019)
10. Yazdi, M.B., Jahed-Motlagh, M.R., Attia, S.A., Raisch, J.: Modal exact linearization of a class of second-order switched nonlinear systems. Nonlinear Anal. Real World Appl. **11**(4), 2243–2252 (2010)
11. Goebel, R., Sanfelice, R.G., Teel, A.R.: Invariance principles for switching systems via hybrid systems techniques. Syst. Control Lett. **57**(12), 980–986 (2008)
12. Lian, J., Shi, P., Feng, Z.: Passivity and passification for a class of uncertain switched stochastic time-delay systems. IEEE Trans. Cybern. **43**(1), 3–13 (2013)
13. Yang, H., Jiang, B., Cocquempot, V., Zhang, H.: Stabilization of switched nonlinear systems with all unstable modes: application to multi-agent systems. IEEE Trans. Autom. Control **56**(9), 2230–2235 (2011)
14. Yang, H., Jiang, B., Zhang, H.: Stabilization of non-minimum phase switched nonlinear systems with application to multi-agent systems. Syst. Control Lett. **61**(10), 1023–1031 (2012)

15. Zhao, X., Zhang, L., Shi, P., Liu, M.: Stability of switched positive linear systems with average dwell time switching. Automatica **48**(6), 1132–1137 (2012)
16. Ordonez-Hurtado, R.H., Duarte-Mermoud, M.A.: Finding common quadratic lyapunov functions for switched linear systems using particle swarm optimisation. Int. J. Control **85**(1), 12–25 (2012)
17. Sun, C., Fang, B., Huang, W.: Existence of a common quadratic lyapunov function for discrete switched linear systems with m stable subsystems. IET Control Theory Appl. **5**(3), 535–537 (2011)
18. Sun, C., Fang, B., Huang, W.: Existence of a common quadratic lyapunov function for discrete switched linear systems with m stable subsystems. IET Control Theory Appl. **5**(3), 535–537 (2011)
19. Sun, Y., Wang, L.: On stability of a class of switched nonlinear systems. Automatica **49**(1), 305–307 (2013)
20. Vu, L., Liberzon, D.: Common lyapunov functions for families of commuting nonlinear systems. Syst. Control Lett. **54**(5), 405–416 (2005)
21. Wu, J.-L.: Stabilizing controllers design for switched nonlinear systems in strict-feedback form. Automatica **45**(4), 1092–1096 (2009)
22. Ma, R., Zhao, J.: Backstepping design for global stabilization of switched nonlinear systems in lower triangular form under arbitrary switchings. Automatica **46**(11), 1819–1823 (2010)
23. Han, T.-T., Ge, S.S., Lee, T.H.: Adaptive neural control for a class of switched nonlinear systems. Syst. Control Lett. **58**(2), 109–118 (2009)
24. Yu, L., Zhang, M., Fei, S.: Non-linear adaptive sliding mode switching control with average dwell-time. Int. J. Syst. Sci. **44**(3), 471–478 (2011)
25. Peleties, P., DeCarlo, R.: Asymptotic stability of m-switched systems using lyapunov-like functions. In: Proceedings of the American Control Conference, pp. 1679–1684 (1991)
26. Ling, H., Michel, A.N., Hui, Y.: Stability analysis of switched systems. In: Proceedings of the 35th IEEE Conference on Decision and Control, vol. 2, pp. 1208–1212
27. Morse, A.S.: Supervisory control of families of linear set-point controllers part i. exact matching. IEEE Trans. Autom. Control **41**(10), 1413–1431 (1996)
28. Hespanha, J.P., Morse, A.S.: Stability of switched systems with average dwell-time. In: Proceedings of the 38th IEEE Conference on Decision and Control, vol. 3, pp. 2655–2660. IEEE
29. Zhai, G., Hu, B., Yasuda, K., Michel, A.N.: Stability analysis of switched systems with stable and unstable subsystems: an average dwell time approach. Int. J. Syst. Sci. **32**, 1055–1061 (2001)
30. Geromel, J.C., Colaneri, P.: H_∞ and dwell time specifications of continuous-time switched linear systems. IEEE Trans. Autom. Control **55**(1), 207–212 (2010)
31. Hespanha, J.P.: Logic-Based Switching Algorithms in Control. Thesis (1998)
32. Sun, X.-M., Liu, G.-P., Wang, W., Rees, D.: L_2-gain of systems with input delays and controller temporary failure: zero-order hold model. IEEE Trans. Control Syst. Technol. **19**(3), 699–706 (2011)
33. Sun, X.-M., Zhao, J., Hill, D.J.: Stability and L_2-gain analysis for switched delay systems: a delay-dependent method. Automatica **42**(10), 1769–1774 (2006)
34. Zhang, L., Gao, H.: Asynchronously switched control of switched linear systems with average dwell time. Automatica **46**(5), 953–958 (2010)
35. Yuan, C., Wu, F.: Hybrid control for switched linear systems with average dwell time. IEEE Trans. Autom. Control **60**(1), 240–245 (2015)
36. Zhai, G., Hu, B., Yasuda, K., Michel, A.N.: Disturbance attenuation properties of time-controlled switched systems. J. Frankl. Inst. **338**(7), 765–779 (2001)
37. Zhao, X., Liu, H., Wang, Z.: Weighted H_∞ performance analysis of switched linear systems with mode-dependent average dwell time. Int. J. Syst. Sci. **44**(11), 2130–2139 (2013)
38. Pettersson, S.: Synthesis of switched linear systems. In: 42nd IEEE International Conference on Decision and Control, vol. 5, pp. 5283–5288
39. Pettersson, S., Lennartson, B.: Stabilization of hybrid systems using a min-projection strategy. Proc. Am. Control Conf. **1**, 223–228 (2001)

40. Ma, R., An, S., Fu, J.: Dwell-time-based stabilization of switched positive systems with only unstable subsystems. Sci. China Inf. Sci. https://doi.org/10.1007/s11432-018-9787-9
41. Feng, S., Wang, J., Zhao, J.: Stability and robust stability of switched positive linear systems with all modes unstable. IEEE/CAA J. Automatica Sinica **6**(1), 167–176 (2019)
42. Liberzon, D., Tempo, R.: Common lyapunov functions and gradient algorithms. IEEE Trans. Autom. Control **49**(6), 990–994 (2004)
43. Mojica-Nava, E., Quijano, N., Rakoto-Ravalontsalama, N., Gauthier, A.: A polynomial approach for stability analysis of switched systems. Syst. Control Lett. **59**(2), 98–104 (2010)
44. Huang, X., Lin, W., Yang, B.: Global finite-time stabilization of a class of uncertain nonlinear systems. Automatica **41**(5), 881–888 (2005)
45. Zhang, X., Feng, G., Sun, Y.: Finite-time stabilization by state feedback control for a class of time-varying nonlinear systems. Automatica **48**(3), 499–504 (2012)
46. Khoo, S., Yin, J., Man, Z., Yu, X.: Finite-time stabilization of stochastic nonlinear systems in strict-feedback form. Automatica **49**(5), 1403–1410 (2013)
47. Bhat, S.P., Bernstein, D.S.: Finite-time stability of continuous autonomous systems. SIAM J. Control Optim. **38**(3), 751–766 (2000)
48. Bhat, S.P., Bernstein, D.S.: Continuous finite-time stabilization of the translational and rotational double integrators. IEEE Trans. Autom. Control **43**(5), 678–682 (1998)
49. Chen, G., Yang, Y., Li, J.: Finite time stability of a class of hybrid dynamical systems. IET Control Theory Appl. **6**(1), 8–13 (2012)
50. Du, H., Li, S., Qian, C.: Finite-time attitude tracking control of spacecraft with application to attitude synchronization. IEEE Trans. Autom. Control **56**(11), 2711–2717 (2011)
51. Menard, T., Moulay, E., Perruquetti, W.: A global high-gain finite-time observer. IEEE Trans. Autom. Control **55**(6), 1500–1506 (2010)
52. Moulay, E., Dambrine, M., Yeganefar, N., Perruquetti, W.: Finite-time stability and stabilization of time-delay systems. Syst. Control Lett. **57**(7), 561–566 (2008)
53. Moulay, E., Perruquetti, W.: Finite time stability and stabilization of a class of continuous systems. J. Math. Anal. Appl. **323**(2), 1430–1443 (2006)
54. Moulay, E., Perruquetti, W.: Finite time stability conditions for non-autonomous continuous systems. Int. J. Control **81**(5), 797–803 (2008)
55. Nersesov, S.G., Nataraj, C., Avis, J.M.: Design of finite-time stabilizing controllers for nonlinear dynamical systems. Int. J. Robust Nonlinear Control **19**(8), 900–918 (2009)
56. Orlov, Y.: Finite time stability and robust control synthesis of uncertain switched systems. SIAM J. Control Optim. **43**(4), 1253–1271 (2005)
57. Polyakov, A.: Nonlinear feedback design for fixed-time stabilization of linear control systems. IEEE Trans. Autom. Control **57**(8), 2106–2110 (2012)
58. Shen, Y., Huang, Y.: Uniformly observable and globally lipschitzian nonlinear systems admit global finite-time observers. IEEE Trans. Autom. Control **54**(11), 2621–2625 (2009)
59. Shen, Y., Huang, Y., Gu, J.: Global finite-time observers for lipschitz nonlinear systems. IEEE Trans. Autom. Control **56**(2), 418–424 (2011)
60. Shen, Y., Xia, X.: Semi-global finite-time observers for nonlinear systems. Automatica **44**(12), 3152–3156 (2008)
61. Seo, S., Shim, H., Seo, J.H.: Global finite-time stabilization of a nonlinear system using dynamic exponent scaling. In: Proceedings of the 47th IEEE Conference on Decision and Control, pp. 3805–3810, Cancun, Mexico (2008)
62. Hong, Y., Wang, J., Cheng, D.: Adaptive finite-time control of nonlinear systems with parametric uncertainty. IEEE Trans. Autom. Control **51**(5), 858–862 (2006)
63. Du, H., Qian, C., Yang, S., Li, S.: Recursive design of finite-time convergent observers for a class of time-varying nonlinear systems. Automatica **49**(2), 601–609 (2013)
64. Hong, Y.: Finite-time stabilization and stabilizability of a class of controllable systems. Syst. Control Lett. **46**(4), 231–236 (2002)
65. Back, J., Cheong, S.G., Shim, H., Seo, J.H.: Nonsmooth feedback stabilizer for strict-feedback nonlinear systems that may not be linearizable at the origin. Syst. Control Lett. **56**(11-12), 742–752, (2007)

66. Ding, S., Li, S.: Global finite-time stabilization of nonlinear integrator systems subject to input saturation. ACTA Automatica Sinica **37**(10), 1222–1231 (2011)
67. Li, J., Qian, C.: Global finite-time stabilization by dynamic output feedback for a class of continuous nonlinear systems. IEEE Trans. Autom. Control **51**(5), 879–884 (2006)
68. Sun, Z., Liu, Y.: Adaptive state-feedback stabilization for a class of high-order nonlinear uncertain systems. Automatica **43**(10), 1772–1783 (2007)
69. Qian, C., Lin, W.: Non-lipschitz continuous stabilizers for nonlinear systems with uncontrollable unstable linearization. Syst. Control Lett. **42**(3), 185–200 (2001)
70. Ding, S., Li, S., Zheng, W.X.: Nonsmooth stabilization of a class of nonlinear cascaded systems. Automatica **48**(10), 2597–2606 (2012)
71. Wang, X., Hong, Y., Jiang, Z.: Finite-time input-to-state stability and optimization of switched systems. In: Proceedings of the 27th Chinese Control Conference, pp. 479–483, 09–11 Dec 2008
72. Han, T.-T., Ge, S.S., Lee, H.T.: Adaptive neural control for a class of switched nonlinear systems. Syst. Control Lett. **58**(2), 109–118 (2009)
73. Long, L., Zhao, J.: H_∞ control of switched nonlinear systems in p-normal form using multiple lyapunov functions. IEEE Trans. Autom. Control **57**(5), 1285–1291 (2012)
74. Ma, R., Liu, Y., Zhao, S., Wang, M., Zong, G.: Nonlinear adaptive control for power integrator triangular systems by switching linear controllers. Int. J. Robust Nonlinear Control **25**(14), 2443–2460 (2015)
75. Long, L., Zhao, J.: Global stabilisation of switched nonlinear systems in p-normal form with mixed odd and even powers. Int. J. Control **84**(10), 1612–1626 (2011)
76. Ma, R., Liu, Y., Zhao, S., Wang, M., Zong, G.: Global stabilization design for switched power integrator triangular systems with different powers. Nonlinear Anal. Hybrid Syst. **15**(2), 74–85 (2015)
77. Long, L., Zhao, J.: Global stabilization of switched nonlinear systems in non-triangular form and its application. J. Frankl. Inst. **365**(2), 1161–1178 (2014)
78. Fu, J., Chai, T., Su, C., Xie, W.: Adaptive output tracking control of a class of nonlinear systems. Control Eng. China **22**(4), 731–736 (2015)
79. Wang, X., Zhao, J.: Logic-based reset adaptation design for improving transient performance of nonlinear systems. IEEE/CAA J. Automatica Sinica **2**(4), 440–448 (2015)
80. Kojic, A., Annaswamy, A.M., Loh, A.P., Lozano, R.: Adaptive control of a class of nonlinear systems with convex/concave parametrization. Syst. Control Lett. **37**(5), 267–274 (1999)
81. Loh, A.P., Annaswamy, A.M., Skantze, F.P.: Adaptation in the presence of general nonlinear parametrization: an error model approach. IEEE Trans. Autom. Control **44**(9), 1634–1652 (1999)
82. Krstic, M., Kanellakopoulos, I., Kokotovic, P.: Nonlinear and Adaptive Control Design. Wiley-Interscience, New York (1995)
83. Lin, W., Qian, C.: Adaptive control of nonlinearly parameterized systems: a nonsmooth feedback framework. IEEE Trans. Autom. Control **47**(5), 757–774 (2002)
84. Chiang, M.-L., Fu, L.-C.: Adaptive stabilization of a class of uncertain switched nonlinear systems with backstepping control. Automatica **50**(8), 2128–2135 (2014)
85. Man, Y., Liu, Y.: Global adaptive stabilisation for nonlinear systems with unknown control directions and input disturbance. Int. J. Control 1–9 (2015)
86. Ye, X.: Global adaptive control of nonlinearly parametrized systems. IEEE Trans. Autom. Control **48**(1), 169–173 (2003)
87. Ye, X.: Nonlinear adaptive control by switching linear controllers. Syst. Control Lett. **61**(4), 617–621 (2012)
88. Sun, Z.Y., Xue, L.R., Zhang, K.: A new approach to finite-time adaptive stabilization of high-order uncertain nonlinear system. Automatica **58**, 60–66 (2015)
89. Wu, J., Chen, W., Li, J.: Global finite-time adaptive stabilization for nonlinear systems with multiple unknown control directions. Automatica **69**, 298–307 (2016)
90. Huang, S., Xiang, Z.: Adaptive finite-time stabilization of a class of switched nonlinear systems using neural networks. Neurocomputing **173**, Part 3, 2055–2061 (2016)

91. Lan, W., Chen, B.: Explicit construction of h_∞ control law for a class of nonminimum phase nonlinear systems. Automatica **44**(3), 738–744 (2008)
92. Qian, C., Lin, W.: Almost disturbance decoupling for a class of high-order nonlinear systems. IEEE Trans. Autom. Control **45**(6), 1208–1214 (2000)
93. Gumussoy, S., Özbay, H.: Sensitivity minimization by strongly stabilizing controllers for a class of unstable time-delay systems. IEEE Trans. Autom. Control **54**(3), 590 (2009)
94. Khargonekar, P.P., Petersen, I.R., Zhou, K.: Robust stabilization of uncertain linear systems: quadratic stabilizability and h_∞ control theory. IEEE Trans. Autom. Control **35**(3), 356–361 (1990)
95. Chai, T.: An indirect stochastic adaptive scheme with on-line choice of weighting polynomials. IEEE Trans. Autom. Control **35**(1), 82–85 (1990)
96. Yue, D., Han, Q.-L., Lam, J.: Network-based robust h_∞ control of systems with uncertainty. Automatica **41**(6), 999–1007 (2005)
97. Fujita, M., Kawai, H., Spong, M.W.: Passivity-based dynamic visual feedback control for three-dimensional target tracking: stability and L_2-gain performance analysis. IEEE Trans. Control Syst. Technol. **15**(1), 40–52 (2007)
98. Allerhand, L.I., Shaked, U.: Robust state-dependent switching of linear systems with dwell time. IEEE Trans. Autom. Control **58**(4), 994–1001 (2013)
99. Duan, C., Wu, F.: Analysis and control of switched linear systems via dwell-time min-switching. Syst. Control Lett. **70**, 8–16 (2014)
100. Hajiahmadi, M., De Schutter, B., Hellendoorn, H.: Stabilization and robust H_∞ control for sector-bounded switched nonlinear systems. Automatica **50**(10), 2726–2731
101. Xu, H., Teo, K.L.: Exponential stability with L_2-gain condition of nonlinear impulsive switched systems. IEEE Trans. Autom. Control **55**(10), 2429–2433 (2010)
102. Briat, C.: Convex lifted conditions for robust l_2-stability analysis and l_2-stabilization of linear discrete-time switched systems with minimum dwell-time constraint. Automatica **50**(3), 976–983 (2014)
103. Briat, C., Seuret, A.: Affine characterizations of minimal and mode-dependent dwell-times for uncertain linear switched systems. IEEE Trans. Autom. Control **58**(5), 1304–1310 (2013)
104. Ma, R., Fu, J., Chai, T.: Dwell-time-based observers design for unknown inputs switched linear systems without requiring strong detectability of subsystems. IEEE Trans. Autom. Control **62**(8), 4215–4221 (2017)
105. Santarelli, K.R., Megretski, A., Dahleh, M.A.: Stabilizability of two-dimensional linear systems via switched output feedback. Syst. Control Lett. **57**(3), 228–235 (2008)
106. Geromel, J.C., Colaneri, P., Bolzern, P.: Dynamic output feedback control of switched linear systems. IEEE Trans. Autom. Control **53**(3), 720–733 (2008)
107. Alessandri, A., Coletta, P.: Switching observers for continuous-time and discrete-time linear systems. In: Proceedings of the 2001 American Control Conference, vol. 3, pp. 2516–2521 (2001)
108. Tanwani, A., Shim, H., Liberzon, D.: Observability for switched linear systems: characterization and observer design. IEEE Trans. Autom. Control **58**(4), 891–904 (2013)
109. Wu, J., Sun, Z.: Observer-driven switching stabilization of switched linear systems. Automatica **49**(8), 2556–2560 (2013)
110. Xie, G., Wang, L.: Periodic stabilizability of switched linear control systems. Automatica **45**(9), 2141–2148 (2009)
111. Yang, J., Chen, Y., Zhu, F., Yu, K., Bu, X.: Synchronous switching observer for nonlinear switched systems with minimum dwell time constraint. J. Frankl. Inst. **352**(11), 4665–4681 (2015)
112. Zhao, X., Liu, H., Zhang, J., Li, H.: Multiple-mode observer design for a class of switched linear systems. IEEE Trans. Autom. Sci. Eng. **12**(1), 272–280 (2015)
113. Defoort, M., Van Gorp, J., Djemai, M., Veluvolu, K.: Hybrid observer for switched linear systems with unknown inputs. In: 7th IEEE Conference on Industrial Electronics and Applications, pp. 594–599 (2012)

114. Huang, G.-J., Chen, W.-H.: A revisit to the design of switched observers for switched linear systems with unknown inputs. Int. J. Control Autom. Syst. **12**(5), 954–962 (2014)
115. Bejarano, F.J., Pisano, A.: Switched observers for switched linear systems with unknown inputs. IEEE Trans. Autom. Control **56**(3), 681–686 (2011)
116. Yang, J., Zhu, F., Tan, X., Wang, Y.: Robust full-order and reduced-order observers for a class of uncertain switched systems. J. Dyn. Syst. Meas. Control **138**(2) (2016)
117. Chen, J., Li, J., Yang, S., Deng, F.: Weighted optimization-based distributed kalman filter for nonlinear target tracking in collaborative sensor networks. IEEE Trans. Cybern. https://doi.org/10.1109/TCYB.2016.2587723
118. Bhat, S.P., Bernstein, D.S.: Finite-time stability of continuous autonomous systems. SIAM J. Control Optim. **38**(3), 751–766 (2000)
119. Lin, W., Qian, C.: Adding one power integrator: a tool for global stabilization of high-order lower-triangular systems. Syst. Control Lett. **39**(5), 339–351 (2000)
120. Haddad, M.M., Nersesov, S.G., Liang, D.: Finite-time stability for time-varying nonlinear dynamical systems. In: 2008 American Control Conference, pp. 4135–4139 (2008)

Chapter 2
Global Finite-Time Stabilization of Switched Nonlinear Systems in Lower-Triangular Form

This chapter deals with the global finite-time stabilization problem for a class of switched nonlinear systems under arbitrary switchings. All subsystems of the studied switched system under consideration are in lower triangular form. Based on the adding one power integrator technique, both a class of non-Lipschitz continuous state feedback controller and a common Lyapunov function are simultaneously constructed such that the closed-loop switched system is global finite-time stability under arbitrary switchings. In the controller design process, a common coordinate transformation of all subsystems is exploited to avoid using individual coordinate transformation for each subsystem. Finally, Two examples are given to show the effectiveness of the proposed method.

2.1 Introduction

As an important class of hybrid dynamical systems, the switched systems have become a hot topic due to the significance both in theory and applications [1, 2]. Considerable effort has been mainly devoted to the analysis and synthesis of switched systems in the last twenty years [3–12]. Many methods, such as, common Lyapunov function [13], multiple Lyapunov functions [14], average dwell-time method [15–17] and so on, have been proposed to solve the stability and stabilization problem for switched systems.

Stability under arbitrary switchings is a very desirable property of switched systems. The existence of a common Lyapunov function for all subsystems was shown to be necessary and sufficient conditions for asymptotically stability of a switched system under arbitrary switchings in [18]. A number of conditions have been made toward the existence of a common Lyapunov function guaranteeing the asymptoti-

J. Fu and R. Ma, *Stabilization and H_∞ Control of Switched Dynamic Systems*, Studies in Systems, Decision and Control 310,
https://doi.org/10.1007/978-3-030-54197-2_2

cally stability under arbitrary switchings [18]. Even though some progress has been made in, such as, [12, 19], finding a common quadratic Lyapunov function for a family of subsystems, especially for switched nonlinear systems, is still an open problem unless they are in some particular form. For instance, [20] presented constructions of a common Lyapunov function for a finite family of pairwise commuting globally asymptotically stable nonlinear systems. In the recent years, switched nonlinear systems in lower triangular form has also drawn considerable attention to study the global stabilization under arbitrary switchings [21, 22] and some designed switching signal [23, 24].

On the other hand, the finite-time control has been widely used in engineering. Finite-time stabilization is one of the most important problems of finite-time control and has been studied in, for example, ([25–27] and the references cited therein). In the recent years, the finite-time control problem for switched nonlinear systems has attracted increasing attention because the finite time control usually demonstrate some desired features. With the help of homogeneous techniques, [28] investigated finite-time stability and robust stability of a class of uncertain switched nonlinear systems. [29] studied the finite-time stability of a class of hybrid dynamical systems with both switching and impulsive effect. Finite-time input-to-state stability problem for switched nonlinear systems was presented in [30]. Although these results provides some methods for studying the finite-time control for switched nonlinear systems, to the authors' best knowledge, few works has been done for the finite-time stabilization problem for switched nonlinear systems based on constructive design method. Therefore, this chapter aims to address this problem.

In this chapter, we will present a construct method to design a non-Lipschitz state feedback control law which globally finite-time stabilizes a class of switched nonlinear systems in lower triangular form under arbitrary switchings. Such a class of switched systems has attracted a lot of attention in the recent years. It is worth pointing out that although some results on global stabilization under arbitrary switchings for the studied switched systems in this chapter have been reported in [21, 22] by backstepping, to the authors' best knowledge, no results are available in literature on finite-time stabilization for such system. This is because the non-Lipschitz continuous function should be suggested in designing the common stabilizing function, which yields a common coordinate transformation for all subsystems during the iterate process. Hence, such a common Lyapunov function can be constructed via the common stabilizing function. So, how to design such a stabilizing function is of great significance, which is solved in this chapter by using the adding a power integrator technique to present a systematic approach of constructing such a stabilizing function. Then, a recursive design algorithm is developed to construct a global finite-time stabilization controller as well as a common Lyapunov function. Compared to the relevant existing results in the literature, this chapter has its owns characteristics. First, we concern the switched nonlinear systems are in lower triangular where the uncertainties appear in the control channel, which covers more general cases. Second, a sufficient condition on global finite-time stabilization is derived and a recursive design algorithm for constructing a finite-time stabilization controller is presented. In addition, a common coordinate transformation at the each step of the adding a power integrator technique is employed to avoid using individual coordinate transformations for each subsystem of switched systems.

2.2 Problem Formulation

We consider the following switched nonlinear systems that, after a suitable change of coordinates, can be expressed in the following form:

$$\begin{aligned} \dot{x}_1 &= x_2 + f_{1,\sigma(t)}(x, u, t), \\ &\cdots \\ \dot{x}_{n-1} &= x_n + f_{n-1,\sigma(t)}(x, u, t), \\ \dot{x}_n &= d(t)u + f_{n,\sigma(t)}(x, u, t), \end{aligned} \tag{2.1}$$

where $x = (x_1, x_2, \ldots, x_n)^T \in R^n$ is the state, $\bar{x}_i = (x_1, \ldots, x_i) \in R^i, i = 1, 2, \ldots, n$, the function $\sigma(t) : [0, \infty) \to M = \{1, \ldots, m\}$ is the switching signal which is assumed to be a piecewise constant or piecewise continuous (from the right) function depending on time or state or both, m is the number of models (called subsystems) of the switched system. For each $k \in M$, $u_k \in R$ is the control input of the kth subsystem. For any $k \in M$, and $i = 1, 2, \ldots, n$, the functions $f_{i,k}(\bar{x}_i) : R^i \to R$, are C^1 with respect to their arguments and vanish at the origin. $d(t)$ is an unknown continuous disturbance and/or parameter belonging to a known compact set $\Omega \in R^s$ and $d(t)$ is away from zero, i.e., $d(t) > 0$. Without loss of generality, we assume that there exist constants d_1 and d_2 such that $d_2 \geq d(t) \geq d_1 > 0$.

In this chapter, we assume that such a switching function $\sigma(t)$ has a finite number of discontinuities on every bounded time interval, and takes a constant value on every interval between two consecutive switching times. We also assume that the state of the switched system (2.1) does not jump at the switching instants, i.e., the trajectory $x(t)$ is everywhere continuous.

Remark 2.1 The switched system (2.1) is more general than many models of the existing results, see, e.g., [21–24]. The advantage of the system under consideration is that the uncertainties appear in the channel, which is more realistic. For non-switched nonlinear systems, this structure has been extensively studied (see, for example, in [25] and references therein).

Now, we give a full characterization of the switched system (2.1) via the following assumption.

Assumption 2.1 There exist known C^1 functions $\mu_{i,k}(\bar{x}_i) \geq 0, i = 1, \ldots, n, \forall k \in M$, such that

$$\left| f_{i,k}(x, u, t) \right| \leq (|x_1| + \cdots + |x_i|)\mu_{i,k}(\bar{x}_i). \tag{2.2}$$

Remark 2.2 Assumption 2.1 is not conservative. In fact, even for non-switched nonlinear system, Assumption 2.1 is quite standard in the literatures of global finite-time stabilization in [25]. In addition, due to the complexity of switched nonlinear systems, this assumption is also reasonable.

The following definition introduces the notion of finite-time stabilization for switched systems (2.1) under arbitrary switchings.

Definition 2.3 The switched system (2.1) is global finite-time stabilization under arbitrary switchings if there is, if possible, a control law such that the switched system (2.1) under arbitrary switchings satisfies the following conditions:

1. Lyapunov-stable: for $\forall \varepsilon > 0$, there exists a $\delta(\varepsilon) > 0$ such that for every $x_0 \in B_\delta(0)$, $x(t) \in B_\varepsilon(0)$ for all $t \geq 0$;
2. Finite time convergence: $\forall x_0 \in R^n$, if there are $t \in [0, T_x(x_0)]$, such that $x(t, x_0) \to 0$ as $t \to T_x(x_0)$.

Remark 2.4 It is easy to see that the finite-time stability of the closed-loop switched nonlinear systems can be regarded as a generalization of the one of the non-switched system (1.3) in Definition 1.1. Note that if the switched system (2.1) is finite time stable, then it is asymptotically stable, and hence, finite time stability is a stronger notion than asymptotic stability.

This chapter aims to design non-Lipschitz continuous state feedback controller by adding a power integrator to globally finite-time stabilize the switched system (2.1) under arbitrary switchings.

2.3 Main Result

The purpose of this chapter is to show that the global finite-time stabilization design under arbitrary switchings for the switched system (2.1) can be solved by a non-Lipschitz continuous state feedback controller via common Lyapunov function method. In addition, both one state feedback control law and a common Lyapunov function can be explicitly constructed by adding a power integrator technique. In the following, we first construct a common Lyapunov function and a controller for the switched system (2.1) and then give the stability analysis.

Step1. Consider the following collection of first order subsystems in switched systems (2.1):

$$\dot{x}_1 = x_2 + f_{1,k}(x, u, t), k = 1, \ldots, m. \tag{2.3}$$

Define $V_1(x_1) = \frac{1}{2}x_1^2$ and set $c = \frac{4n}{2n+1}$.

With the help of Assumption 2.1, differentiating $V_1(x_1)$ along the trajectories of all subsystems in (2.3) gives

$$\begin{aligned}\dot{V}_1(x_1) &= x_1(x_2 + f_{1,k}(x,u,t)) \\ &\le x_1x_2 + x_1^2\mu_{1,k}(x_1) \\ &\le x_1(x_2 - x_2^*) + x_1x_2^* + x_1^c v_{1,k}, \forall k \in M, \end{aligned} \tag{2.4}$$

where a C^1 function $v_{1,k}(x_1) \ge x_1^{2/(2n+1)}\mu_{1,k}(x_1) \ge 0$.

Taking x_2 as the virtual control in (2.3), there exists a continuous feedback controller

$$x_2^* = -x_1^{\rho_2}\varphi_1(x_1), \tag{2.5}$$

where a C^1 function $\varphi_1(x_1) \ge n + \max_{k\in M}\{v_{1,k}(x_1)\}$ and $\rho_2 = (2n-1)/(2n+1)$.

Similar to [22], such a function x_2^* is called common stabilization function in this chapter. Note that $\varphi_1(x_1)$ is a function independent of k. Thus, the virtual control x_2^* is also a function independent of k, which will be used to yield a common coordinate transformation of all subsystems in the next step of applying the adding a power integrator technique.

Then, substituting (2.4) into (2.5) yields

$$\dot{V}_1(x_1) \le -nx_1^c + x_1(x_2 - x_2^*). \tag{2.6}$$

Inductive step: For the $(i-1)$th step, we will use induction. We assume that, for the following switched system

$$\dot{x}_j = x_{j+1} + f_{j,\sigma(t)}(x,u,t), \quad j = 1, 2, \ldots, i-1, \tag{2.7}$$

there is a C^1, positive definite and proper function $V_{i-1}(\bar{x}_{i-1})$ satisfying

$$V_{i-1}(\bar{x}_{i-1}) \le 2(\chi_1^2 + \cdots + \chi_{i-1}^2), \tag{2.8}$$

where the common stabilization functions and the common coordinate transformations are defined by

$$\begin{aligned} &x_1^* = 0, && \chi_1 = x_1^{1/\rho_1} - x_1^{*1/\rho_1}, \\ &x_2^* = -\chi_1^{\rho_2}\varphi_1(x_1), && \chi_2 = x_2^{1/\rho_2} - x_2^{*1/\rho_2}, \\ &\cdots && \cdots \\ &x_i^* = -\chi_{i-1}^{\rho_i}\varphi_{i-1}(\bar{x}_{i-1}), && \chi_i = x_i^{1/\rho_i} - x_i^{*1/\rho_i} \end{aligned} \tag{2.9}$$

with $\rho_1 = 1 > \rho_2 > \cdots > \rho_i \triangleq \frac{2n+3-2i}{2n+1} > 0$, and $\varphi_1(x_1) > 0, \cdots, \varphi_{i-1}(\bar{x}_{i-1}) > 0$ being C^1 function, such that, for all subsystems of the switched systems (2.7),

$$\dot{V}_{i-1} \le \chi_{i-1}^{2-\rho_{i-1}}(x_i - x_i^*) - (n-i+2)\sum_{l=1}^{i-1}\chi_l^c. \tag{2.10}$$

In the following, our goal is to construct $V_i(\bar{x}_i)$ and a common stabilization x_{i+1}^* such that the analog of (2.8), (2.9) and (2.10) holds with i replacing $i-1$.

Define

$$V_i(\bar{x}_i) = V_{i-1}(\bar{x}_{i-1}) + \Gamma_i(\bar{x}_i), \tag{2.11}$$

where

$$\Gamma_i(\bar{x}_i) = \int_{x_i^*}^{x_i} (\lambda^{1/\rho_i} - x_i^{*1/\rho_i})^{2-\rho_i} d\lambda.$$

First, we introduce four propositions for the sake of following process.

Proposition 2.5 *Function $\Gamma_i(\bar{x}_i)$ is C^1. Besides, $\frac{\partial \Gamma_i}{\partial x_i} = \chi_i^{2-\rho_i}$ and*

$$\frac{\partial \Gamma_i}{\partial x_l} = -(2-\rho_i)\frac{\partial (x_i^{*1/\rho_i})}{\partial x_l} \int_{x_i^*}^{x_i} (\lambda^{1/\rho_i} - x_i^{*1/\rho_i})^{1-\rho_i} d\lambda,$$
$$l = 1, \ldots, i-1. \tag{2.12}$$

Proposition 2.6 *Function $V_i(\bar{x}_i)$ is a C^1, positive definite and proper function, and satisfies*

$$V_i(\bar{x}_i) \le 2(\chi_1^2 + \cdots \chi_i^2). \tag{2.13}$$

Proposition 2.7 *There exist C^1 functions $\tilde{\mu}_{i,k}(\bar{x}_i) \ge 0$, $i = 1, \ldots, n$, and $\forall k \in M$, satisfying*

$$\left| f_{i,k}(x, u, t) \right| \le (|\chi_1|^{\rho_i} + \cdots + |\chi_i|^{\rho_i})\tilde{\mu}_{i,k}(\bar{x}_i). \tag{2.14}$$

Proposition 2.8 *There exist nonnegative C^1 functions $\Psi_{i,l,k}(\bar{x}_i)$, $l = 1, \ldots, i-1$, satisfying*

$$\left| \frac{\partial (x_i^{*1/\rho_i})}{\partial x_l} \dot{x}_l \right| \le \Psi_{i,l,k}(\bar{x}_i) \sum_{j=1}^{i} \left| \chi_j \right|^{(2n-1)/(2n+1)}. \tag{2.15}$$

Form [25], the proofs of Propositions 2.5–2.8 are straightforward and hence left to the reader as an exercise.

With the help of Proposition 2.5, differentiating $V_i(\bar{x}_i)$ along all the trajectories of all subsystems in (2.1), we obtain that

$$\dot{V}_i \le -(n-i+2)\sum_{l=1}^{i-1} \chi_l^c + \chi_{i-1}^{2-\rho_{i-1}}(x_i - x_i^*) + \chi_i^{2-\rho_i}(x_{i+1} - x_{i+1}^*)$$
$$+\chi_i^{2-\rho_i} x_{i+1}^* + \chi_i^{2-\rho_i} f_{i,k} + \sum_{l=1}^{i-1} \frac{\partial \Gamma_i}{\partial x_l}\dot{x}_l,\ k = 1, \ldots, m. \tag{2.16}$$

In the following, we will estimate the upper bound of each term on the right hand of (2.16).

Using Lemma 1.4 and $\rho_i = \rho_{i-1} - 2/(2n+1)$, we have

$$\left|x_i - x_i^*\right| \le 2^{1-\rho_i} \left|x_i^{1/\rho_i} - (x_k^*)^{1/\rho_i}\right|^{\rho_i} \le 2\left|\chi_i\right|^{\rho_i},$$

$$\left|\chi_{i-1}^{2-\rho_{i-1}}(x_i - x_i^*)\right| \le 2\left|\chi_{i-1}\right|^{2-\rho_{i-1}} \left|\chi_i\right|^{\rho_i} \le \frac{\chi_{i-1}^c}{3} + \varepsilon_i \chi_i^c, \tag{2.17}$$

where $\varepsilon_i > 0$ is a fixed constant.

Using Lemma 1.3 and Proposition 2.7, for each $k \in M$, the following inequalities hold:

$$\begin{aligned}\left|\chi_i^{2-\rho_i} f_{i,k}(x,u,t)\right| &\le \left|\chi_i\right|^{2-\rho_i} \sum_{\tau=1}^{i} \left|\chi_\tau\right|^{\rho_i - 2/(2n+1)} \tilde{\mu}_{i,k}(\bar{x}_i) \\ &\le \frac{1}{3}\sum_{\tau=1}^{i-1} \chi_\tau^c + \chi_i^c \tilde{\zeta}_{i,k}(\bar{x}_i),\end{aligned} \tag{2.18}$$

where C^1 function $\tilde{\mu}_{i,k}(\bar{x}_i) \ge 0$, $\tilde{\zeta}_{i,k}(\bar{x}_i) \ge 0$, $\forall \bar{x}_i \in R^i$, $\forall k \in M$.

Now an application of Proposition 2.5 and 2.8 yields that

$$\begin{aligned}\left|\sum_{l=1}^{i-1} \frac{\partial \Gamma_i}{\partial x_l}\dot{x}_l\right| &\le 2(2-\rho_i)\left|\chi_i\right| \sum_{l=1}^{i} \left|\chi_l\right|^{(2n-1)/(2n+1)} \sum_{l=1}^{i-1} \Psi_{i,l,k}(\bar{x}_i) \\ &\le \frac{1}{3}\sum_{l=1}^{i-1} \chi_l^c + \chi_i^c \hat{\zeta}_{i,k}(\bar{x}_i),\end{aligned} \tag{2.19}$$

where $\hat{\zeta}_{i,k}(\bar{x}_i) \ge 0$ is a C^1 function.

Substituting (2.17), (2.18) and (2.19) into (2.16) yields

$$\begin{aligned}\dot{V}_i \le &-(n-i+1)\sum_{l=1}^{i-1} \chi_l^c + \chi_i^{2-\rho_i}(x_{i+1} - x_{i+1}^*) + \chi_i^{2-\rho_i} x_{i+1}^* \\ &+ \chi_i^c \left[\varepsilon_i + \tilde{\zeta}_{i,k}(\bar{x}_i) + \hat{\zeta}_{i,k}(\bar{x}_i)\right].\end{aligned} \tag{2.20}$$

We choose the common stabilization function

$$x_{i+1}^* = -\chi_i^{\rho_{i+1}} \varphi_i(\bar{x}_i)$$

with a C^1 function $\varphi_i(\bar{x}_i) \ge n - i + 1 + \varepsilon_i + \max_{k\in M}\{\tilde{\zeta}_{i,k}(\bar{x}_i) + \hat{\zeta}_{i,k}(\bar{x}_i)\}$ and $0 < \rho_{i+1} \triangleq \rho_i - 2/(2n+1) < \rho_i$.

It is easy for us to get

$$\dot{V}_i \leq -(n-i+1)\sum_{l=1}^{i} \chi_l^c + \chi_i^{2-\rho_i}(x_{i+1} - x_{i+1}^*), \forall k \in M. \tag{2.21}$$

Now, we achieve the proof of the inductive step.

Step n: Using repeatedly the inductive argument above, it is straightforward to see that at the last step (Step n), one can explicitly construct a change of coordinates of the form (2.9), a C^1 positive definite and proper Lyapunov function

$$V_n(x) = V_{n-1}(\bar{x}_{n-1}) + \Gamma_n(x), \tag{2.22}$$

where $\Gamma_n(x) = \int_{x_n^*}^{x_n} (\lambda^{1/\rho_n} - x_i^{*1/\rho_n})^{2-\rho_n} d\lambda$, satisfying

$$V_n(x) \leq 2(\chi_1^2 + \cdots \chi_n^2) \tag{2.23}$$

such that

$$\begin{aligned} \dot{V}_n \leq & -\sum_{l=1}^{n-1} \chi_l^c + \chi_n^{2-\rho_n}\left(d(t)x_{n+1} - d_1 x_{n+1}^*\right) + d_1 \chi_n^{2-\rho_n} x_{n+1}^* \\ & + \chi_n^c\left(\varepsilon_n + \tilde{\zeta}_n(x) + \hat{\zeta}_n(x)\right). \end{aligned} \tag{2.24}$$

Then, we choose a static smooth state feedback control law

$$\begin{aligned} u = x_{n+1}^* = & -\frac{1}{d_1}\chi_n^{\rho_n - 2/(2n+1)}\left(1 + \varepsilon_n + \tilde{\zeta}_n(x) + \hat{\zeta}_n(x)\right) \\ \stackrel{\Delta}{=} & -\frac{1}{d_1}\chi_n^{1/(2n+1)}\varphi_n(x), \end{aligned} \tag{2.25}$$

which yields

$$\dot{V}_n \leq= -\sum_{l=1}^{n} \chi_l^c - \chi_n^c \frac{d(t) - d_1}{d_1}\varphi_n(x) \leq -\sum_{l=1}^{n} \chi_l^c. \tag{2.26}$$

That is,

$$\dot{V}_n(x) \leq -\left[\chi_1^{4n/(2n+1)} + \cdots + \chi_n^{4n/(2n+1)}\right]. \tag{2.27}$$

Taking $\alpha = \frac{2n}{2n+1} \in (0, 1)$ and using Lemma 1.4, we have

$$V_n^\alpha(x) \leq 2\sum_{l=1}^{n} \chi_l^{4n/(2n+1)}. \tag{2.28}$$

Moreover, we can obtain the inequality:

$$\dot{V}_n(x) + \frac{1}{4}V_n^{\alpha}(x) \le -\frac{1}{2}\sum_{l=1}^{n} \chi_l^{4n/(2n+1)} \le 0 \tag{2.29}$$

for each subsystem $k \in M$, and settling time satisfies $T_x(x_0) \le \frac{4V(x_0)^{1-\alpha}}{(1-\alpha)}$.

We summarize our main result as follows.

Theorem 2.9 *The close-loop switched system (2.1) with (2.25) is globally finite-time stable under arbitrary switchings with common Lyapunov function $V_n(x)$. In addition, the settling time satisfies $T_x(x_0) \le \frac{4V(x_0)^{1-\alpha}}{1-\alpha}$.*

Proof It follows directly form (2.27) and (2.29).

2.4 Simulation Studies

In this section, two practical examples are studied to show the effectiveness of the proposed method. One is the CSTR with two modes feed stream and the other is the haptic display system with switched virtual environments, both of which will show potential applications of the proposed method.

Example 2.10 In this section, we use the obtained method to the CSTR with two modes feed stream, which is molded as a switched system and some issues of this switched system have been studied in [31, 32] and so on.

As shown in [22], under a coordinate transformation and smooth feedback, such a CSTR with two modes feed stream can be changed into the following form:

$$\begin{aligned} \dot{x}_1 &= x_2 + f_{1,\sigma(t)}(x_1) \\ \dot{x}_2 &= u \end{aligned} \tag{2.30}$$

where $\sigma(t) \to M = \{1, 2\}$, $f_{1,1}(x_1) = 0.5x_1$ and $f_{1,2}(x_1) = 2x_1$.

For such a switched system (2.30), the global stabilization problem under arbitrary switchings has been studied in [22]. Here, we will design a controller to achieve the global finite-time stabilization problem under arbitrary switchings, which is a stronger property than the global asymptotical stabilization in [22].

First, consider the x_1-equation of each subsystem and view x_2 as the input. We define

$$V_1(x_1) = \frac{1}{2}x_1^2. \tag{2.31}$$

By Step 1 in the previous section, we can show that

$$x_2^* = -3x_1^{3/5}(2 + 2x_1^{2/5})$$

is a common stabilization function. In this case, we have

$$\dot{V}_1(x_1) \leq -2x_1^{8/5} + x_1\left(x_2 - x_2^*\right). \tag{2.32}$$

Define

$$V_2(x) = V_1(x_1) + \int_{x_2^*}^{x_2} \left(\lambda^{5/3} - x_2^{*5/3}\right)^{2-3/5} d\lambda \tag{2.33}$$

as the candidate common Lyapunov function for both subsystems in (2.30).

By (2.25), we can obtain the finite-time stabilizing controller

$$u = -(x_1^{5/3} - x_2^{*5/3})^{1/5} \times (3 + 3(60x_1 + 50x_2 + 6x_2^2 + 10x_1^2)), \tag{2.34}$$

which results in

$$\dot{V}_2(x) \leq -\left(x_1^{8/5} + (x_2^{5/3} + x_2^{*5/3})^{8/5}\right), \tag{2.35}$$

and further obtains

$$\dot{V}_2(x) + \frac{1}{4}V_2^{4/5}(x) \leq -\frac{1}{2}\left(x_1^{8/5} + (x_2^{5/3} + x_2^{*5/3})^{8/5}\right) \leq 0. \tag{2.36}$$

Then, according to Theorem 2.9, the closed-loop switched system (2.30) with (2.34) is global finite-time stable under arbitrary switchings.

Let $x_1(0) = -2.5$, $x_2(0) = 3.8$. Figure 2.1 shows the state trajectories of the closed-loop switched system (2.30) with (2.34) under some randomly chosen switching signal shown in Fig. 2.2. It can be seen that the closed-loop switched system is finite-time stable.

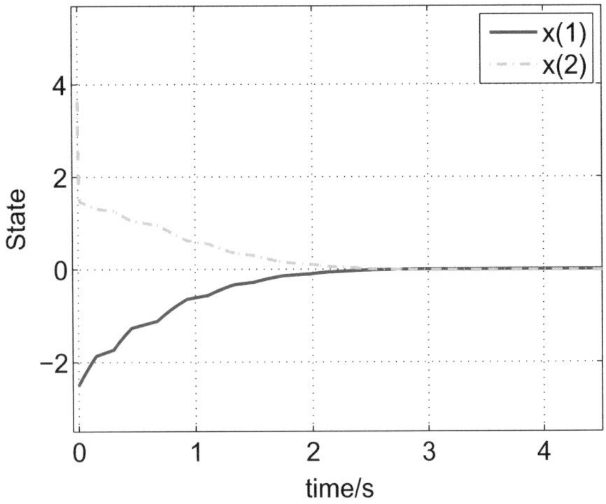

Fig. 2.1 State responses of the system (2.30)–(2.34)

Fig. 2.2 The switching signal

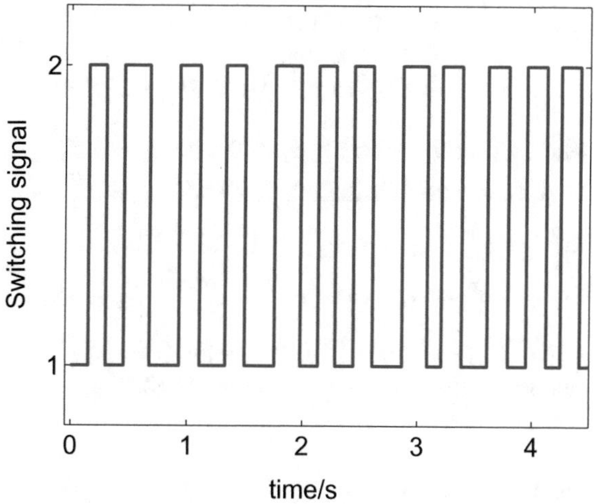

Fig. 2.3 State responses of the system (2.30) with $u = -4(x_2 + 3x_1)$

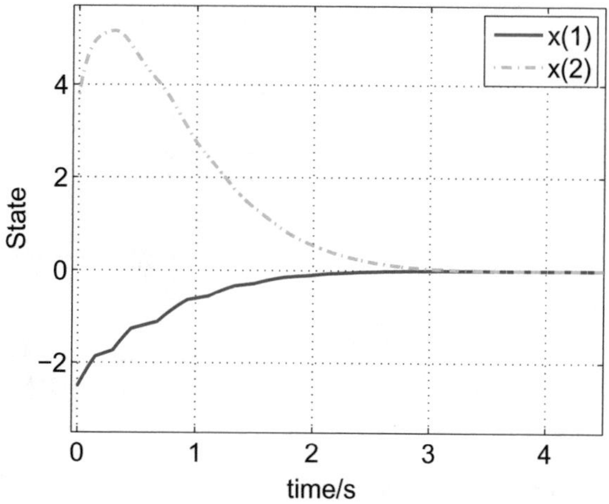

It is worth pointing out that the finite-time stabilizer (2.34) is very different to the asymptotic stabilizer $u = -4(x_2 + 3x_1)$ proposed in [22]. Figure 2.3 shows the simulation result of controller designed by [22] under the same initial state and the switching signal described in Fig 2.2. By comparing Fig. 2.1 and Fig. 2.3, one can see that the convergence rate of state responses under our proposed finite-time stabilizer is faster than the one under the controller designed by [22], which obviously shows one of the advantages of the finite time stabilization control.

Example 2.11 As shown in [33, 34], after some manipulations some haptic display system with switched virtual environments can be modeled as the switched system (2.1) with the following subsystems:

$$\begin{array}{ll} \dot{x}_1 = -12x_1 + x_2, & \dot{x}_1 = -12x_1 + x_2, \\ \dot{x}_2 = -x_2 + x_3 & \dot{x}_2 = -0.5x_2 + x_3, \\ \dot{x}_3 = 50x_2 - 0.5x_3 + d(t)u, & \dot{x}_3 = 20x_2 - x_3 + d(t)u, \end{array} \tag{2.37}$$

where $d(t) : d_2 \geq d(t) \geq d_1 > 0$ with $d_1 = 1.5, d_2 = 2.5$.

It is easy to obtain that $x_2^* = -3x_1^{5/7}, \chi_2 = x_2^{7/5} - x_2^{*7/5}, \varepsilon_2 = \frac{5}{12}\left(\frac{4}{7}\right)^{-\frac{7}{5}}, \tilde{\mu}_2 = 3 + 3x_1^{2/7} + \chi_2^{2/7}$, $\hat{\zeta}_2 = \frac{7}{12}\left(\frac{5}{4}\right)^{\frac{5}{7}}\left[\frac{18}{7}\left(15 + 12x_1^{2/7}\right)\right]^{12/7}$, $\tilde{\zeta}_2 = \frac{3}{4}\left(\frac{4}{3}\tilde{\mu}_2^{-4}\right)^{-\frac{1}{3}} + \tilde{\mu}_2$, $\varphi_2 = 2 + \varepsilon_2 + \tilde{\zeta}_2 + \hat{\zeta}_2$, $\chi_3 = x_3^{7/3} - x_3^{*7/3}$, $\tilde{\mu}_3 = 150\left(1 + x_1^{4/7} + \chi_2{}^{4/7}\right) + \varphi_2$, $\varepsilon_3 = \frac{1}{4}\left(\frac{4}{9}\right)^{-9/5}, \tilde{\zeta}_3 = \frac{3}{2}\left(\frac{4}{3}\right)^{-\frac{3}{11}}\tilde{\mu}_3^{12/11} + \tilde{\mu}_3, \hat{\zeta}_3 = \frac{7}{6}\left(\frac{4}{5}\right)^{-5/7}\left(\frac{22}{7}\Psi_3\right)^{12/7} + \frac{22}{7}\psi_3$. Finally, we can obtain the following controller by (2.25):

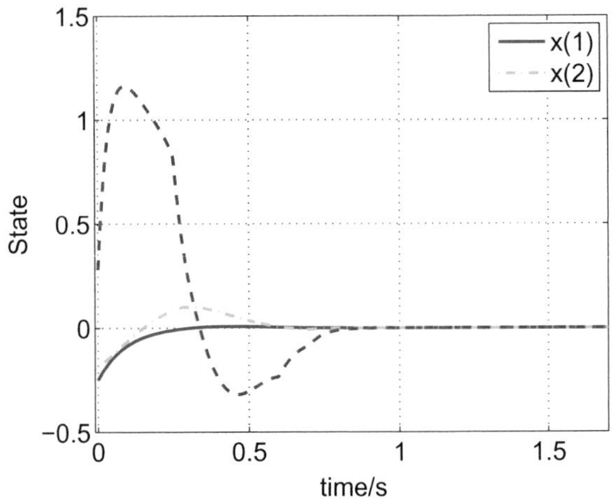

Fig. 2.4 State responses of the switched system (2.37)

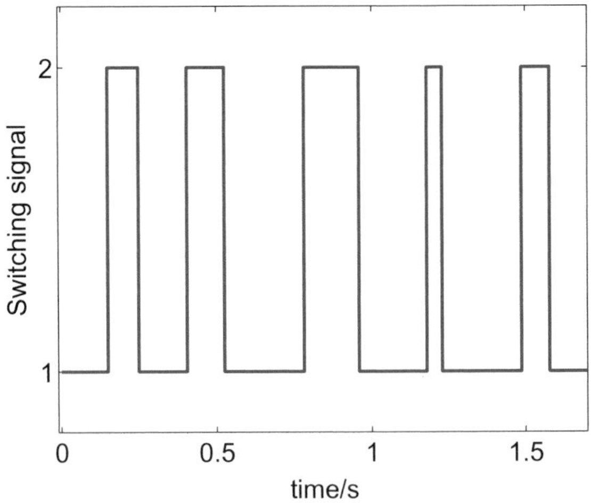

Fig. 2.5 The switching signal

$$u = -\frac{2}{3}\chi_3^{1/7}\left[1 + \varepsilon_3 + \tilde{\zeta}_3 + \hat{\zeta}_3\right]. \tag{2.38}$$

Figures 2.4 and 2.5 show the trajectories of the states and the switching signal of the closed-loop switched system (2.1) with subsystems (2.37), respectively. The switching signal is also generated randomly. From Fig. 2.5, it can be seen that the states of the closed-loop system also converge to zero in finite time.

2.5 Conclusions

In this chapter, we have investigated the global finite-time stabilization problem for a class of switched nonlinear systems in lower triangular form under arbitrary switchings by adding one power integrator technique. Both a common state feedback controller independent of switching signals and a common Lyapunov function are constructed to global finite-time stabilize the switched system under arbitrary switchings. Construction of a common stabilizing function plays a crucial role in the method and the underlying idea behind our control design is to adopt a feedback domination approach to control design. Two application examples has been employed to show the effectiveness of the proposed finite-time control design method.

References

1. Ibanez, C.A., Suarez-Castanon, M.S., Gutierrez-Frias, O.O.: A switching controller for the stabilization of the damping inverted pendulum cart system. Int. J. Innov. Comput. Inf. Control **9**(9), 3585–3597 (2013)
2. Wang, X., Zhao, J.: Switched adaptive tracking control of robot manipulators with friction and changing loads. Int. J. Syst. Sci. (ahead-of-print), 1–11 (2013)
3. Hu, Q., Cheng, D.: Stabilizer design of planar switched linear systems. Syst. Control Lett. **57**(10), 876–879 (2008)
4. Wu, Z.-G., Shi, P., Su, H., Chu, J.: Stochastic synchronization of Markovian jump neural networks with time-varying delay using sampled data. IEEE Trans. Cybern. **43**(6), 1796–1806 (2013)
5. Ma, R., Zhao, J., Dimirovski, G.M.: Backstepping design for global robust stabilisation of switched nonlinear systems in lower triangular form. Int. J. Syst. Sci. **44**(4), 615–624 (2013)
6. Wu, L., Su, X., Shi, P.: Sliding mode control with bounded l-2 gain performance of Markovian jump singular time-delay systems. Automatica **48**(8), 1929–1933 (2012)
7. Wu, Z.-G., Shi, P., Su, H., Chu, J.: Delay-dependent stability analysis for switched neural networks with time-varying delay. IEEE Trans. Syst. Man Cybern. Part B: Cybern. **41**(6), 1522–1530 (2011)
8. Liu, J., Liu, X., Xie, W.-C.: Delay-dependent robust control for uncertain switched systems with time-delay. Nonlinear Anal. Hybrid Syst. **2**(1), 81–95 (2008)
9. Xiang, Z., Xiang, W.: Stability analysis of switched systems under dynamical dwell time control approach. Int. J. Syst. Sci. **40**(4), 347–355 (2009)
10. Su, X., Shi, P., Wu, L., Song, Y.-D.: A novel control design on discrete-time takagi-sugeno fuzzy systems with time-varying delays. IEEE Trans. Fuzzy Syst. **21**(4), 655–671 (2013)

11. Wu, L., Su, X., Shi, P.: Output feedback control of Markovian jump repeated scalar nonlinear systems. IEEE Trans. Autom. Control (2013)
12. Sun, Y., Wang, L.: On stability of a class of switched nonlinear systems. Automatica **49**(1), 305–307 (2013)
13. Ordonez-Hurtado, R.H., Duarte-Mermoud, M.A.: Finding common quadratic lyapunov functions for switched linear systems using particle swarm optimisation. Int. J. Control **85**(1), 12–25 (2012)
14. Branicky, M.S.: Multiple lyapunov functions and other analysis tools for switched and hybrid systems. IEEE Trans. Autom. Control **43**(4), 475–482 (1998)
15. Zhang, J., Han, Z., Zhu, F., Huang, J.: Stability and stabilization of positive switched systems with mode-dependent average dwell time. Nonlinear Anal. Hybrid Syst. **9**, 42–55 (2013)
16. Zhai, G., Hu, B., Yasuda, K., Michel, A.N.: Stability analysis of switched systems with stable and unstable subsystems: an average dwell time approach. Int. J. Syst. Sci. **32**, 1055–1061 (2001)
17. Zhang, L., Gao, H.: Asynchronously switched control of switched linear systems with average dwell time. Automatica **46**(5), 953–958 (2010)
18. Liberzon, Daniel: Switching in Systems and Control. Birkhauser, Boston (2003)
19. Sun, C., Fang, B., Huang, W.: Existence of a common quadratic lyapunov function for discrete switched linear systems with m stable subsystems. IET Control Theory Appl. **5**(3), 535–537 (2011)
20. Vu, L., Liberzon, D.: Common lyapunov functions for families of commuting nonlinear systems. Syst. Control Lett. **54**(5), 405–416 (2005)
21. Wu, J.-L.: Stabilizing controllers design for switched nonlinear systems in strict-feedback form. Automatica **45**(4), 1092–1096 (2009)
22. Ma, R., Zhao, J.: Backstepping design for global stabilization of switched nonlinear systems in lower triangular form under arbitrary switchings. Automatica **46**(11), 1819–1823 (2010)
23. Han, T.-T., Ge, S.S., Lee, T.H.: Adaptive neural control for a class of switched nonlinear systems. Syst. Control Lett. **58**(2), 109–118 (2009)
24. Yu, L., Zhang, M., Fei, S.: Non-linear adaptive sliding mode switching control with average dwell-time. Int. J. Syst. Sci. **44**(3), 471–478 (2011)
25. Huang, X., Lin, W., Yang, B.: Global finite-time stabilization of a class of uncertain nonlinear systems. Automatica **41**(5), 881–888 (2005)
26. Zhang, X., Feng, G., Sun, Y.: Finite-time stabilization by state feedback control for a class of time-varying nonlinear systems. Automatica **48**(3), 499–504 (2012)
27. Khoo, S., Yin, J., Man, Z., Yu, X.: Finite-time stabilization of stochastic nonlinear systems in strict-feedback form. Automatica **49**(5), 1403–1410 (2013)
28. Orlov, Y.: Finite time stability and robust control synthesis of uncertain switched systems. SIAM J. Control Optim. **43**(4), 1253–1271 (2005)
29. Chen, G., Yang, Y., Li, J.: Finite time stability of a class of hybrid dynamical systems. IET Control Theory Appl. **6**(1), 8–13 (2012)
30. Wang, X., Hong, Y., Jiang, Z.: Finite-time input-to-state stability and optimization of switched systems. In: Proceedings of the 27th Chinese Control Conference, pp. 479–483, 09–11 Dec 2008
31. Yazdi, M.B., Jahed-Motlagh, M.R., Attia, S.A., Raisch, J.: Modal exact linearization of a class of second-order switched nonlinear systems. Nonlinear Anal. Real World Appl. **11**(4), 2243–2252 (2010)
32. Mhaskar, P., El-Farra, N.H., Christofides, P.D.: Predictive control of switched nonlinear systems with scheduled mode transitions. IEEE Trans. Autom. Control **50**(11), 1670–1680 (2005)
33. Jin, Y., Fu, J., Jing, Y.: Fault-tolerant control of a class of switched systems with strong structural uncertainties with application to haptic display systems. Neurocomputing **103**, 143–148 (2013)
34. Mahapatra, S., Zefran, M.: Stable haptic interaction with switched virtual environments. In: Proceedings of the 2003 IEEE International Conference on Robotics and Automation, vol. 1, pp. 1241–1246. IEEE (2003)

Chapter 3
Global Finite-Time Stabilization of Switched Nonlinear Systems with the Powers of Positive Odd Rational Numbers

In this chapter, we will present new results on global finite-time stabilization for a class of switched strict-feedback nonlinear systems, whose subsystems have chained integrators with the powers of positive odd rational numbers (i.e., numerators and denominators of the powers are all positive odd integers but not necessarily relatively prime). All the powers in each equation of subsystems of the switched systems can be different. Based on the technique of adding a power integrator, the global finite-time stabilizers of individual subsystems are first systematically constructed to guarantee global finite-time stability of the closed-loop smooth switched system under arbitrary switchings, and then a co-design of stabilizers and a state-dependent switching law is proposed to achieve global finite-time stabilization of the closed-loop non-smooth switched systems. In the controller design, a common coordinate transformation of all subsystems is exploited to avoid using individual coordinate transformations for individual subsystems. We also give some sufficient conditions that enable our design by characterizing the powers of the chained integrators of the considered switched systems. Numerical examples are provided to demonstrate the effectiveness of the proposed results.

3.1 Introduction

In previous chapter, the condition on global finite-time stabilization of switched nonlinear systems in lower-triangular form are obtained. One feature of the studied switched systems in previous chapter is that the powers of the chained integrators are restricted to 1. However, in practical applications, the powers of the chained integrators may no longer be required to be 1. In the past few years, the switched nonlinear systems in p-normal form has received much attention and a few important

J. Fu and R. Ma, *Stabilization and H_∞ Control of Switched Dynamic Systems*, Studies in Systems, Decision and Control 310,
https://doi.org/10.1007/978-3-030-54197-2_3

results have also appeared in, for example, [1, 2]. One feature of the studied switched systems in [1] is that the powers of the chained integrators are restricted to the same positive odd integer for subsystems during the whole operation time, and thus these methods are not applicable to switched systems where all powers in each equation of subsystems of the switched systems can be different. The authors of [3] studied the more general switched systems in p-normal form, where the powers of the subsystems are allowed to be different and to be positive even integers, and obtained a sufficient condition under which a globally asymptotic stabilizer is designed. The authors of [2] investigated the global exponential stabilization under arbitrary switchings for a class of switched nonlinear systems in power integrator triangular form, whose subsystems have chained integrators with the powers of different positive odd numbers. It should be noted that the powers of the chained integrators in [2, 3] are restricted to positive integers.

In this chapter, we will focus on the case where the powers of the chained integrators (of switched nonlinear systems) are positive odd rational numbers (i.e., numerators and denominators of the powers are all positive odd integers but not necessarily relatively prime such as $\frac{3}{5}$, $\frac{7}{3}$ and $\frac{15}{5}$). It is worth pointing out that no results even on the *asymptotic* stabilization are available in the literature. However, considering the desired properties of finite-time stabilization, we shall study the finite-time stabilization of switched strict-feedback nonlinear systems in p-normal form where the chained integrators have the powers of positive odd rational numbers, and where powers of the chained integrators can be different for each subsystem during the whole operation time. Usually, when the power of the chained integrators are different, we have to construct a common Lyapunov function under different coordinate transformation, which may be tedious and difficult to apply. Therefore, obtaining a common coordinate transformation for all subsystems during iterations is desired, which is also a crucial issue for the technique of adding a power integrator to be applicable to switched nonlinear systems. This has not been studied in the existing literature, which also partially motivates our present work.

Specifically, first a sufficient condition on the relationship between the powers of subsystems of the switched systems for its global finite-time stabilization is derived, then with the technique of adding a power integrator, a recursive design algorithm is developed to construct a globally finite-time stabilizer as well as a C^1 positive definite and proper common Lyapunov function, and last a co-design of finite-time stabilizers and a state-dependent switching law is proposed to achieve global finite-time stabilization of the closed-loop non-smooth switched systems. Compared to the relevant existing results in the literature, the main contributions of this chapter are as follows.

- The global finite-time stabilization problem of the switched nonlinear systems, which does not have to have first approximation at the origin and where the powers of the chained integrators of each equation can be different, is studied for the first time.
- A sufficient condition on characterizing the powers of the chained integrators of the switched nonlinear systems for its global finite-time stabilization is derived.

- A recursive design algorithm for constructing a global finite-time stabilizer of the switched nonlinear system is presented, and a common coordinate transformation at each step of the technique of adding a power integrator is used to avoid using individual coordinate transformations for individual subsystems of switched systems.
- Another feature of the paper in designing the common Lyapunov function is the selection of the powers since the powers should be appropriately selected to reflect the future design steps to come in the technique of adding a power integrator.
- The last contribution is the co-design of stabilizers and a switching policy proposed to achieve its finite-time stabilization of the switched nonlinear systems also containing non-smooth nonlinearities.

This chapter is organized as follows. Section 3.2 introduces preliminaries and the system description. A systematic design procedure is presented in Sect. 3.3. In Sect. 3.4, some important special cases of the main result of Sect. 3.3 are discussed in detail. Based on the previous development, the co-design of stabilizers and a switching law is proposed in Sect. 3.5. Illustrative design examples are given in Sect. 3.6. Finally, some conclusions are drawn in Sect. 3.7.

3.2 Preliminaries and System Description

We consider the following switched nonlinear systems:

$$\begin{aligned}
\dot{x}_1 &= x_2^{r_{1,\sigma(t)}} + f_{1,\sigma(t)}(x_1),\\
\dot{x}_2 &= x_3^{r_{2,\sigma(t)}} + f_{2,\sigma(t)}(x_1, x_2),\\
&\cdots\\
\dot{x}_n &= u_{\sigma(t)}^{r_{n,\sigma(t)}} + f_{n,\sigma(t)}(x_1, x_2, \ldots, x_n),
\end{aligned} \tag{3.1}$$

where $x = (x_1, x_2, \ldots, x_n) \in R^n$ is the state; $\bar{x}_i = (x_1, \ldots, x_i) \in R^i, i = 1, 2, \ldots, n$; the function $\sigma(t) : [0, \infty) \to M = \{1, \ldots, m\}$ is the switching signal which is assumed to be a piecewise constant or piecewise continuous (from the right) function depending on time or state or both; m is the number of models (called subsystems) of the switched system. For each $k \in M$, $u_k \in R$ is the control input of the kth subsystem. For any $k \in M$, and $i = 1, 2, \ldots, n$, the functions $f_{i,k}(\bar{x}_i) : R^i \to R$, are C^1 with respect to their arguments and vanish at the origin, and $r_{i,k} \in Q_{odd}$. In addition, we assume that the state of the switched system (3.1) does not jump at the switching instants, i.e., the trajectory $x(t)$ is everywhere continuous.

Remark 3.1 In switched systems, jumps in the solution depends on the reset map (also called reset matrix in [4] for linear reset map) which resets the state of the active subsystem at the switching instants. The reset map determines whether the state at the switching instants jumps. If the reset map is an identity map, the state of the switched system is continuous (i.e., no jumps). If at some switching instant the reset map resets the state of switched systems to a certain value (say x^*), then the state of the just activated subsystem will evolve starting from this new reset value x^*. Many published papers assumed that the state of switched systems does not jump at the switching instants (i.e., the state of switched systems is everywhere continuous), such as in [5–9]. There are some papers focusing on the switched systems with state jumps, which are also referred to switched impulsive systems [4, 10]. Most recently, there is an interesting paper focusing on a class of switched systems where part of states are continuous but the rest of states jumps [11]. As big part of switched system literature did [5–9], in our paper we also made this commonly-used assumption of state continuity at switching instants.

Remark 3.2 Note that, for non-switched nonlinear systems, that is, $m = 1$, the typical structure of system (3.1) has been studied for its globally finite-time stabilization in [12, 13]. On the other hand, when $r_{i,k} = 1$ for all $k \in M$, $i = 1, 2, \ldots, n$, the switched system (3.1) reduces to the well-known switched systems in strict-feedback form whose global stabilization problem has been addressed in [14–16]. So, the structure of the switched system (3.1) is much more general than those of the switched systems in [14–16].

Our control objective is to design a state feedback controller to globally finite-time stabilize the system (3.1) under arbitrary switchings. This will be done by simultaneously constructing both the state feedback controller and a C^1 common Lyapunov function for all subsystems based on the technique of adding a power integrator, such that all subsystems of the closed-loop switched system (3.1) satisfy Lemma 1.6. Furthermore, another objective is to co-design stabilizers and a switching law to finite-time stabilize system (3.1) containing non-smooth nonlinearities.

Note that if the closed-loop system (3.1) is finite-time stable, then it is also asymptotically stable. Hence, finite-time stability is a stronger notion than asymptotic stability.

To achieve the proposed control objective, we need the following assumptions.

Assumption 3.1 For given numbers $r_{i,k}, k \in M, i = 1, 2, \ldots, n$, in (3.1) there exist constants $q_i \geq 1$, $q_i \in Q_{odd}$, $i = 1, 2, \ldots, n+1$, and $\tau_k \in (0, 1)$, $\tau_k \in Q_{odd}$ such that

$$q_1 = 1, \frac{r_{i,k}}{q_{i+1}} + 1 = \tau_k + \frac{1}{q_i}, i = 1, 2, \ldots, n, \tag{3.2}$$

$$\frac{r_{i,k}}{q_{i+1}} \leq \min\left\{\frac{1}{q_1}, \frac{1}{q_2}, \ldots, \frac{1}{q_i}\right\}, i = 1, 2, \ldots, n. \tag{3.3}$$

Assumption 3.2 σ has finite number of switchings on any finite interval of time.

Remark 3.3 Assumption 3.1 is checkable or tractable as shown in Corollaries 3.13–3.16 in Sect. 3.4. Zeno behavior of switched systems (or hybrid systems) occurs if infinitely many switchings or discrete transitions take place in a finite time interval. Zeno behavior is one of important issues in switched systems and hybrid systems, which has been studied to some extent [17]. For example necessary and sufficient condition for the existence of Zeno executions are given in [17], and observability normal form for switched systems with Zeno phenomena [18]. As many switched systems papers did [5, 6, 19], we used Assumption 3.2 to exclude the Zeno behavior.

3.3 Global Finite-Time Stabilization

In this section, we will construct a common Lyapunov function and a finite-time stabilizer for switched systems (3.1), and then give the stability analysis. Both the global finite-time stabilizers and a common Lyapunov function will be explicitly constructed simultaneously with the aid of the technique of adding a power integrator.

In the following, we first present the finite-time controller design step-by-step for switched systems (3.1). The process to construct the global finite-time stabilizers is completed by induction argument, where a common coordinate transformation for all subsystems is exploited.

Before constructing the controller, one can show that under Assumption 3.1 $\frac{r_{i,k}}{q_{i+1}} = \tau_k + \frac{1}{q_i} - 1$ and $q_i \geq 1$, and thus that

$$0 < \frac{r_{i,k}}{q_{i+1}} < 1, \quad i = 1, 2, \ldots, n. \tag{3.4}$$

Since $f_{i,k} : R^i \to R, \forall k \in M$ and $i = 1, 2, \ldots, n$, are C^1 functions with $f_{i,k}(0) = 0$, then, by [20], there exist C^1 nonnegative functions $\gamma_{i,k}(\bar{x}_i)$ such that

$$\left| f_{i,k}(\bar{x}_i) \right| \leq (|x_1| + \cdots + |x_i|)\gamma_{i,k}(\bar{x}_i). \tag{3.5}$$

Step 1. For the first step of the induction, choose a quadratic Lyapunov function $V_1(x_1) = \frac{1}{2}x_1^2$. For each subsystem k of switched systems (3.1), by using (3.5), a simple computation gives

$$\begin{aligned}
\dot{V}_1 &\leq x_1 \left(x_1^{r_{1,k}} - x_2^{*r_{1,k}}\right) + x_1 x_2^{*r_{1,k}} + x_1^2 \gamma_{1,k} \\
&\leq x_1 \left(x_1^{r_{1,k}} - x_2^{*r_{1,k}}\right) + x_1 x_2^{*r_{1,k}} + |x_1|^{1+\frac{r_{1,k}}{q_2}} \left(1 + |x_1|^{1-\frac{r_{1,k}}{q_2}}\right) \gamma_{1,k} \\
&\leq x_1 \left(x_1^{r_{1,k}} - x_2^{*r_{1,k}}\right) + x_1 x_2^{*r_{1,k}} + |x_1|^{1+\frac{r_{1,k}}{q_2}} \tilde{\gamma}_{1,k}(x_1),
\end{aligned} \tag{3.6}$$

where x_2^* is the C^0 virtual controller to be designed later and $\tilde{\gamma}_{1,k}(x_1)$ is a smooth function satisfying

$$\tilde{\gamma}_{1,k}(x_1) \geq (1 + |x_1|^{1-r_{1,k}/q_2})\gamma_{1,k}(x_1), \forall k \in M.$$

For the first virtual control x_2^* of each subsystem $k \in M$, a common state feedback control law is designed as:

$$x_2^* = -\beta_1(x_1)x_1^{1/q_2}, \tag{3.7}$$

with a C^1 function $\beta_1(x_1) \geq \max_{k\in M} \left(n + \tilde{\gamma}_{1,k}(x_1)\right)^{1/r_{1,k}}$. Note that $\beta_1(x_1)$ is a function independent of k. Thus, the virtual control x_2^* is also a function independent of k, which will be used to yield a common coordinate transformation of all subsystems in the next step of applying the adding a power integrator technique.

Substituting (3.7) into (3.6) gives

$$\dot{V}_1(x_1) \leq -nx_1^{1+\tau_k} + x_1\left[x_2^{r_{1,k}} - x_2^{*r_{1,k}}\right]. \tag{3.8}$$

Remark 3.4 It should be noted that x_2^* and $x_2^{*r_{1,k}}$ may not be C^1 over the entire R, while $x_2^{*q_2}$ is C^1 over the entire R due to $q_2 \geq 1$.

Inductive step. For the ith ($i = 2, 3, \ldots, n$) step we use induction. Suppose at step $i-1$, there are a C^1 Lyapunov function $V_{i-1}(\bar{x}_{i-1})$, which is positive definite and proper, satisfying

$$V_{i-1}(\bar{x}_{i-1}) \leq 2\left(\xi_1^2 + \cdots + \xi_{i-1}^2\right), \tag{3.9}$$

and a set of common virtual controllers $x_1^*, x_2^*, \ldots, x_i^*$, which are C^1 except at the origin $\bar{x}_i = 0$ (i.e., C^0 at the origin), defined by

$$\begin{aligned} x_1^* &= 0, & \xi_1 &= x_1 - x_1^*, \\ x_2^* &= -\beta_1(x_1)\xi_1^{\frac{1}{q_2}}, & \xi_2 &= x_2^{q_2} - x_2^{*q_2}, \\ &\vdots & &\vdots \\ x_i^* &= -\beta_{i-1}(\bar{x}_{i-1})\xi_{i-1}^{\frac{1}{q_i}}, & \xi_i &= x_i^{q_i} - x_i^{*q_i}, \end{aligned} \tag{3.10}$$

with a set of C^1 functions $\beta_j(\bar{x}_j) > 0$, $1 \leq j \leq i-1$, such that

$$\begin{aligned} \dot{V}_{i-1}(\bar{x}_{i-1}) \leq{}& -(n-i+2)(\xi_1^{1+\tau_k} + \cdots + \xi_{i-1}^{1+\tau_k}) \\ &+\xi_{i-1}^{2-\frac{1}{q_{i-1}}}(x_i^{r_{i-1,k}} - x_i^{*r_{i-1,k}}). \end{aligned} \tag{3.11}$$

Note that such virtual control laws in (3.10) are independent of k, which guarantees that the coordinate transformations are identical for all subsystems in the subsequent steps of applying the technique of adding one power integrator.

Our goal is to construct the common virtual control law x_{i+1}^* and the common Lyapunov function $V_i(\bar{x}_i)$, and thus we claim that the (3.9) and (3.11) hold at step i. To prove this claim, we consider a Lyapunov function candidate defined by

$$V_i(\bar{x}_i) = V_{i-1}(\bar{x}_{i-1}) + W_i(\bar{x}_i), \tag{3.12}$$

where $W_i(\bar{x}_i) = \int_{x_i^*}^{x_i} (s^{q_i} - x_i^{*q_i})^{2-\frac{1}{q_i}} ds$. Thus, we need to give the following proposition proved in Appendix 3.8 to show that $V_i(\bar{x}_i)$ is C^1, positive definite and proper.

Proposition 3.5 *$V_i(\bar{x}_i)$ is a C^1, positive definite and proper function, and $W_i(\bar{x}_i)$ is C^1 with the properties: for $j = 1, 2, \ldots, i-1$,*

$$\frac{\partial W_i(\bar{x}_i)}{\partial x_j} = -\int_{x_i^*}^{x_i} (2 - \frac{1}{q_i})(s^{q_i} - x_i^{*q_i})^{1-\frac{1}{q_i}} ds \, \frac{\partial (x_i^{*q_i})}{\partial x_j}, \tag{3.13}$$

$$\frac{\partial W_i(\bar{x}_i)}{\partial x_i} = (x_i^{q_i} - x_i^{*q_i})^{2-\frac{1}{q_i}} = \xi_i^{2-\frac{1}{q_i}}. \tag{3.14}$$

We now can continue constructing the controller. Using (3.1) and (3.11)–(3.14), a straightforward calculation gives

$$\begin{aligned}
\dot{V}_i &= \dot{V}_{i-1} + \frac{\partial W_i(\bar{x}_i)}{\partial x_i}\dot{x}_i + \sum_{j=1}^{i-1} \frac{\partial W_i(\bar{x}_i)}{\partial x_j}\dot{x}_j \\
&\leq -(n-i+2)\sum_{j=1}^{i-1} \xi_j^{1+\tau_k} + \xi_{i-1}^{2-\frac{1}{q_{i-1}}} (x_i^{r_{i-1,k}} - x_i^{*r_{i-1,k}}) \\
&\quad + \xi_i^{2-\frac{1}{q_i}} \left[x_{i+1}^{r_{i,k}} + f_{i,k}(\bar{x}_i)\right] + \sum_{j=1}^{i-1} \frac{\partial W_i(\bar{x}_i)}{\partial x_j}\dot{x}_j.
\end{aligned} \tag{3.15}$$

Next, to estimate the last three terms on the right-hand side of (3.15), we introduce the following propositions specifically for system (3.1), whose proofs are given in Appendices 3.8.2, 3.8.3 and 3.8.4.

Proposition 3.6 *There are constants $c_{i,1,k} > 0$, $k = 1, 2, \ldots, m$, such that*

$$\left| \xi_{i-1}^{2-\frac{1}{q_{i-1}}} (x_i^{r_{i-1,k}} - x_i^{*r_{i-1,k}}) \right| \leq \frac{1}{3} \sum_{j=1}^{i-1} \xi_j^{1+\tau_k} + c_{i,1,k}\xi_i^{1+\tau_k}. \tag{3.16}$$

Proposition 3.7 *There are C^1 functions $c_{i,2,k}(\bar{x}_i) > 0$, $k = 1, 2, \ldots, m$, such that*

$$\left| \xi_i^{2-\frac{1}{q_i}} f_{i,k}(\bar{x}_i) \right| \leq \frac{1}{3} \sum_{j=1}^{i-1} \xi_j^{1+\tau_k} + \xi_i^{1+\tau_k} c_{i,2,k}(\bar{x}_i). \tag{3.17}$$

Proposition 3.8 *There exist C^1 functions $c_{i,3,k}(\bar{x}_i) > 0$, $k = 1, 2, \ldots, m$, such that*

$$\left| \sum_{j=1}^{i-1} \frac{\partial W_i(\bar{x}_i)}{\partial x_j}\dot{x}_j \right| \leq \frac{1}{3} \sum_{j=1}^{i-1} \xi_j^{1+\tau_k} + c_{i,3,k}(\bar{x}_i)\xi_i^{1+\tau_k}. \tag{3.18}$$

Now, with the help of Propositions 3.6–3.8, we substitute (3.16), (3.17) and (3.18) into (3.15) and obtain that

$$\dot{V}_i \le -(n-i+1)\sum_{j=1}^{i-1}\xi_j^{1+\tau_k} + \left[c_{i,1,k}+c_{i,2,k}+c_{i,3,k}\right]\xi_i^{1+\tau_k} + \xi_i^{2-\frac{1}{q_i}} x_{i+1}^{*r_{i,k}} + \xi_i^{2-\frac{1}{q_i}}\left(x_{i+1}^{r_{i,k}} - x_{i+1}^{*r_{i,k}}\right). \tag{3.19}$$

By Assumption 3.1, one has $2-\frac{1}{q_i}+\frac{r_{i,k}}{q_{i+1}} = 1+\tau_k$. Then, for the first virtual control x_{i+1}^* of each subsystem $k \in M$, we construct a common state feedback control law as

$$x_{i+1}^* = -\beta_i(\bar{x}_i)\xi_i^{\frac{1}{q_{i+1}}}, \tag{3.20}$$

where $\beta_i(\bar{x}_i)$ is C^1 independent of k and satisfies

$$\beta_i(\bar{x}_i) \ge \max_{k\in M}\left(c_{i,1,k}+c_{i,2,k}+c_{i,3,k}+n-i+1\right)^{\frac{1}{r_{i,k}}}, \tag{3.21}$$

and obtain

$$\dot{V}_i \le -(n-i+1)\sum_{j=1}^{i}\xi_j^{1+\tau_k} + \xi_i^{2-\frac{1}{q_i}}(x_{i+1}^{r_{i,k}} - x_{i+1}^{*r_{i,k}}). \tag{3.22}$$

Inequality (3.22) is the desired ith step of inequality (3.11), which proves the induction.

Note again that x_{i+1}^* and $x_{i+1}^{*r_{i,k}}$ may not be C^1, but $x_{i+1}^{*q_{i+1}}$ is.

Remark 3.9 There always exists a C^1 function $\beta_i(\bar{x}_i)$. In the following, we present a way to choose such a function. For simplicity, define

$$y_k = c_{i,1,k} + c_{i,2,k}(\bar{x}_i) + c_{i,3,k}(\bar{x}_i) + n - i + 1, \forall k \in M.$$

Without loss of generality, we assume that $r_{i,1}, r_{i,2}, \ldots, r_{i,l} \le 1$ with $l \in M$ and $0 \le l \le m$ and other $r_{i,l+1}, r_{i,l+2}, \ldots, r_{i,m} > 1$. For all $r_{i,k} \le 1$, one can choose

$$\beta_{i,1}(\bar{x}_i) = \sum_{k=1}^{r} y_k^{\frac{1}{r_{i,k}}}.$$

On the other hand, for $r_{i,k} > 1$, since $2 - 1/r_{i,k} > 0$ and $1^{2-\frac{1}{r_{i,k}}} y_k^{\frac{1}{r_{i,k}}} \le \frac{2-1/r_{i,k}}{2} + \frac{1}{2r_{i,k}} y_k^2$ by (1.9), then one can select

$$\beta_{i,2}(\bar{x}_i) = \sum_{k=r+1}^{m} \frac{1}{2r_{i,k}} y_k^2 + \max_{k \in M} \left\{ 1 - \frac{1}{2r_{i,k}} \right\}.$$

It is easy to prove that such selected function

$$\beta_i(\bar{x}_i) = \beta_{i,1}(\bar{x}_i) + \beta_{i,2}(\bar{x}_i)$$

is C^1.

Step n. Using the inductive argument above, we conclude that at the nth step, there exists a C^0 state feedback control law of the form

$$u_k = x_{n+1}^* = -\beta_n(x)\xi_n^{\frac{1}{q_{n+1}}}, \ \forall\, k = 1, 2, \ldots, m, \tag{3.23}$$

with the C^1 proper and positive definite Lyapunov function $V_n(x)$ constructed via the inductive procedure, such that

$$\begin{aligned} \dot{V}_n &\leq -\left(\xi_1^{1+\tau_k} + \cdots + \xi_n^{1+\tau_k}\right) + \xi_n^{2-\frac{1}{q_n}} \left(u_k^{r_{n,k}} - x_{n+1}^{*r_{n,k}}\right) \\ &= -\left(\xi_1^{1+\tau_k} + \cdots + \xi_n^{1+\tau_k}\right). \end{aligned} \tag{3.24}$$

Recall that

$$V_n(x) = \sum_{j=1}^{n} W_j(\bar{x}_j), \tag{3.25}$$

where $W_j(\bar{x}_j)$ is defined in (3.12) and $W_1(x_1) = V_1(x_1)$. Then, we obtain

$$V_n^{c_k}(x) = [\sum_{j=1}^{n} W_j(\bar{x}_j)]^{c_k} \leq \sum_{j=1}^{n} W_j^{c_k}(\bar{x}_j), \ \forall x \in R^n. \tag{3.26}$$

Moreover, for $j = 1, 2, \ldots, n$, we have

$$\begin{aligned} W_j(\bar{x}_j) &\leq \left|x_j - x_j^*\right| \left|x_j^{q_j} - x_j^{*q_j}\right|^{2-\frac{1}{q_j}} \\ &= [\left|x_j - x_j^*\right|^{q_j}]^{\frac{1}{q_j}} \left|x_j^{q_j} - x_j^{*q_j}\right|^{2-\frac{1}{q_j}} \\ &\leq \left[2^{q_j-1}\left|x_j^{q_j} - x_j^{*q_j}\right|\right]^{\frac{1}{q_j}} \left|x_j^{q_j} - x_j^{*q_j}\right|^{2-\frac{1}{q_j}} \\ &\leq 2^{\frac{q_j-1}{q_j}} \left|x_j^{q_j} - x_j^{*q_j}\right|^{\frac{1}{q_j}+2-\frac{1}{q_j}} \leq 2\left|\xi_j\right|^2. \end{aligned} \tag{3.27}$$

That is $W_j(\bar{x}_j) \leq 2\left|\xi_j\right|^2$. Thus, we obtain that

$$V_n(x) \leq 2\left(\xi_1^2 + \cdots + \xi_n^2\right). \tag{3.28}$$

By using (3.24), it is easy to see that

$$\begin{aligned}\dot{V}_n &\leq -\sum_{j=1}^{n} \xi_j^{1+\tau_k} \\ &= -\sum_{j=1}^{n} (\xi_j^2)^{\frac{1+\tau_k}{2}} \\ &\leq -(\sum_{j=1}^{n} \xi_j^2)^{\frac{1+\tau_k}{2}} \\ &\leq -(\frac{1}{2}V_n)^{\frac{1+\tau_k}{2}} \\ &= -c_k V_n^{\frac{1+\tau_k}{2}},\end{aligned} \tag{3.29}$$

where $c_k = 2^{-\frac{1+\tau_k}{2}} > 0$. That is,

$$\dot{V}_n(x) + c_k V_n^{\frac{1+\tau_k}{2}}(x) \leq 0. \tag{3.30}$$

It is easy to verify that the power $\frac{1+\tau_k}{2} \in (0, 1)$ due to $\tau_k \in (0, 1)$.

To this end, we are ready to state our main result.

Theorem 3.10 *Consider the switched nonlinear systems (3.1) satisfying Assumptions 3.1 and 3.2, the closed-loop switched systems (3.1) with (3.23) is globally finite-time stable under arbitrary switchings.*

Proof To our Lyapunov function

$$V_n(t, x) \triangleq V_n(x) = \sum_{j=1}^{n} W_j(\bar{x}_j), \tag{3.31}$$

we have

(1) $V_n(t, 0) = 0, t \in [0, +\infty)$;

(2) From Proposition 3.5 we know that V_n is a C^1, positive definite and proper function. Thus, according to Lemma 4.2 of [21], for V_n there exists a class κ function $\alpha(\cdot)$, such that $\alpha(\|x\|) \leq V_n(t, x), t \in [0, \infty), x \in R^n$;

(3) From (3.30) we have

$$\dot{V}_n(t, x) \leq -c_k V_n^{\frac{1+\tau_k}{2}}(t, x), t \in [0, \infty), x \in R^n \tag{3.32}$$

when $\sigma(t) = k, \forall k \in M$.

Therefore, according to finite-time Lyapunov stability theorem for time-varying systems (i.e., Lemma 1.6), V_n is the common Lyapunov function for all subsystems of the closed-loop switched systems (3.1) with (3.23). Thus the proposed controller (3.23) globally finite-time stabilizes the switched systems (3.1) under arbitrary switchings.

Remark 3.11 In Theorem 3.10, f_i represents nonlinearities of the system and vanishes at the origin. Theorem 3.10 cannot cover the cases of unknown parameters and/or unknown structural uncertainties. However, our results can be easily extended to these cases by adding an adaptive law to estimate the unknown parameters (which have to be constants or very slow time-varying parameters) or by using robust control techniques to accommodate unknown structural uncertainties under certain boundedness or growth conditions.

Remark 3.12 Since the technique we used is based on domination, unlike backstepping which is based on cancellation, the magnitude of the control may be supposed to be big. For this issue, our method can be extended to deal with the control saturation problem by integrating bounded control methods in the literature.

3.4 Special Cases

It is worth noting that not all powers of switched nonlinear systems (3.1) satisfy Assumption 3.1. In fact, Assumption 3.1 is not necessarily true even for the subsystem of (3.1). For example, consider the subsystem k of (3.1) with $r_{1,k} = \frac{1}{5}$, $r_{2,k} = 1$, $r_{3,k} = 1$. According to Assumption 3.1, we can calculate that $q_2 = \frac{r_{1,k}}{\tau_k} \geq 1$, $q_3 = \frac{r_{2,k}}{\tau_k + \tau_k / r_{1,k} - 1} \geq 1$, and $q_4 = \frac{r_{3,k}}{\tau_k + (\tau_k + \tau_k / r_{1,k} - 1)/r_{2,k} - 1} \geq 1$, which yields that $\tau_k \leq \frac{1}{5}$, $\frac{2}{6} \geq \tau_k > \frac{1}{6}$, and $\frac{3}{7} \geq \tau_k > \frac{2}{7}$. So there are no parameters τ_k and q_i, $i = 2, 3, 4$ to satisfy Assumption 3.1. In this section, the conditions of Theorem 3.10 will be investigated to design the controller explicitly. In particular, several special cases are provided such that the conditions are satisfied automatically. Although these conditions seem stringent, these conditions are explicitly given to identify specifically what structural properties switched systems (3.1) should have to make it globally finite-time stabilized.

Corollary 3.13 *If the exponents $r_{1,k}$'s in (3.1) with $n = 1$ satisfy $r_{1,k} \in \mathcal{Q}_{odd}$, then, (3.2) and (3.3) follow, and thus, (3.1) admits a globally finite-time stabilizer.*

Proof When $n = 1$ and $q_1 = 1$, from (3.2), we obtain

$$1 + \frac{r_{1,k}}{q_2} = \tau_k + \frac{1}{q_1}.$$

That is,

$$q_2 = \frac{r_{1,k}}{\tau_k}.$$

We can choose a $q_2 > 1$ such that

$$\tau_k = \frac{r_{1,k}}{q_2} \in (0, 1), \forall k \in M.$$

Then, we can verify that (3.2) holds. Keeping (3.2) in mind, we get

$$\frac{r_{1,k}}{q_2} = -1 + \tau_k + \frac{1}{q_1} < \frac{1}{q_1}.$$

Thus (3.3) holds. Then, from Theorem 3.10 a globally finite-time stabilizer for the system (3.1) can be constructed.

Corollary 3.14 *If $r_{1,k} = r_{2,k} = \cdots = r_{n,k} < 1, \forall k \in M$, then, Assumption 3.1 holds and the special case of (3.1) also admits a globally finite-time stabilizer.*

Proof Let $q_1 = q_2 = \cdots = q_n = q_{n+1} = 1$. Then, we choose $\tau_k = r_{1,k} = r_{2,k} = \cdots = r_{n-1,k} = r_{n,k}$, from which one can easily prove that (3.2) holds for each k. In addition, since $r_{1,k} = r_{2,k} = \cdots = r_{n,k} < 1, \forall k \in M$, it is easy to see that (3.3) holds. Therefore, (3.1) also admits a globally finite-time stabilizer.

Corollary 3.15 *If $r_{i,k} \geq 1, i = 1, 2, \ldots, n, \forall k \in M$, the condition (3.3) in Assumption 3.1 can be removed.*

Proof When $i = 1$ and $q_1 = 1$, from Corollary 3.13, the conclusion is correct.

Since $\tau_k - 1 < 0$, one can easily obtain

$$\frac{r_{i,k}}{q_{i+1}} = \tau_k - 1 + \frac{1}{q_i} < \frac{1}{q_i}. \tag{3.33}$$

With the help of (3.2), we have

$$\frac{1}{q_i} = \frac{1}{r_{i-1,k}}\left[\tau_k - 1 + \frac{1}{q_{i-1}}\right], i = 2, \ldots, n. \tag{3.34}$$

Then, with the help of (3.33) and (3.34), we get

$$\frac{r_{i,k}}{q_{i+1}} < \frac{\tau_k - 1}{r_{i-1,k}} + \frac{1}{r_{i-1,k}q_{i-1}} < \frac{1}{r_{i-1,k}}\frac{1}{q_{i-1}} \leq \frac{1}{q_{i-1}} \tag{3.35}$$

by using the fact that $\tau_k - 1 < 0$.

In the following, substituting (3.34) into (3.35) yields

$$\frac{r_{i,k}}{q_{i+1}} < \frac{1}{r_{i-2,k}}(\tau_k - 1 + \frac{1}{q_{i-2}}) < \frac{1}{r_{i-2,k}q_{i-2}} < \frac{1}{q_{i-2}} \tag{3.36}$$

since $\tau_k - 1 < 0$ and $r_{i-2,k} \geq 1$. Then, by applying $\frac{1}{q_i}$ recursively to (3.35) with them, it is easy to show that

$$\frac{r_{i,k}}{q_{i+1}} < \frac{1}{q_{i-2}} < \frac{1}{r_{i-3,k}q_{i-3}} \leq \frac{1}{q_{i-3}} < \cdots$$
$$< \frac{1}{r_{2,k}q_2} \leq \frac{1}{q_2} < \frac{1}{r_{1,k}q_1} \leq \frac{1}{q_1}. \tag{3.37}$$

Then, one can conclude that inequalities (3.3) hold. The proof is then completed.

Next, based on Corollary 3.15, we present another sufficient condition for Assumption 3.1.

Corollary 3.16 *Let $r_{i,k} \geq 1$, $i = 1, 2, \ldots, n$, $\forall k \in M$. If there exist constants τ_k, $\forall k \in M$, such that, for all $j, k \in M$ and $j \neq k$,*

$$\frac{r_{1,k} + r_{1,k}r_{2,k} + \cdots + r_{1,k}r_{2,k}\cdots r_{n-1,k}}{1 + r_{1,k} + r_{1,k}r_{2,k} + \cdots + r_{1,k}r_{2,k}\cdots r_{n-1,k}} < \tau_k <$$
$$\min\left\{1, r_{1,k}, \frac{r_{1,k} + r_{1,k}r_{2,k}}{1 + r_{1,k}}, \ldots, \right.$$
$$\left. \frac{r_{1,k} + r_{1,k}r_{2,k} + \cdots + r_{1,k}r_{2,k}\cdots r_{n,k}}{1 + r_{1,k} + r_{1,k}r_{2,k} + \cdots + r_{1,k}r_{2,k}\cdots r_{n-1,k}}\right\}, \tag{3.38}$$
$$\frac{r_{1,j}}{\tau_j} = \frac{r_{1,k}}{\tau_k}, \tag{3.39}$$
$$\frac{r_{1,j}r_{2,j}\cdots r_{i-2,j}}{(1 + R_{i-2,j})\tau_j - R_{i-2,j}} = \frac{r_{1,k}r_{2,k}\cdots r_{i-2,k}}{(1 + R_{i-2,k})\tau_k - R_{i-2,k}},$$
$$i = 3, \ldots, n+1, \tag{3.40}$$

where $R_{i-2,j} = r_{1,j} + r_{1,j}r_{2,j} + \cdots + \prod_{l=1}^{i-2} r_{l,j}$ with $i = 3, 4, \ldots, n+1$ and $j \in M$. Then, there exist constants q_i, $i = 1, 2, \ldots, n+1$, such that Assumption 3.1 holds.

Proof From Corollary 3.15, we only need to prove that (3.2) holds.

With $q_1 = 1$, $q_2 = \frac{r_{1,j}}{\tau_j} = \frac{r_{1,k}}{\tau_k}$, and $q_i =$

$$\frac{r_{1,j}r_{2,j}\cdots r_{i-2,j}}{(1 + R_{i-2,j})\tau_j - R_{i-2,j}} = \frac{r_{1,k}r_{2,k}\cdots r_{i-2,k}}{(1 + R_{i-2,k})\tau_k - R_{i-2,k}},$$
$$i = 3, 4, \ldots, n+1. \tag{3.41}$$

According to the definition of q_i, $i = 1, 2, \ldots, n+1$, it is easy to verify that (3.2) holds. In the following, we prove that $q_i \geq 1$, $i = 1, 2, \ldots, n+1$ when (3.38) holds.

According to (3.38), we know that for $i = 3, \ldots, n+1$,

$$\begin{aligned}&\frac{r_{1,k}r_{2,k}\cdots r_{i-2,k}}{1+r_{1,k}+\cdots+r_{1,k}r_{2,k}\cdots r_{i-2,k}}\\&<\frac{r_{1,k}r_{2,k}\cdots r_{n-1,k}}{1+r_{1,k}+\cdots+r_{1,k}r_{2,k}\cdots r_{n-1,k}}\\&<\tau_k<\frac{r_{1,k}r_{2,k}\cdots r_{i-1,k}}{1+r_{1,k}+\cdots+r_{1,k}r_{2,k}\cdots r_{i-2,k}}.\end{aligned} \tag{3.42}$$

With the help of the definition of q_i in (3.41) and (3.42), we can directly obtain $q_i \geq 1, i = 3, \ldots, n+1$.

If some powers of system (3.1) *are greater than 1 and other powers are less than 1*, Assumption 3.1 may not hold. The next result shows under certain additional conditions inequality (3.3) in Assumption 3.1 can be removed.

Corollary 3.17 *If $r_{1,k} < 1$, $r_{2,k}r_{3,k} \geq 1$, and $r_{i,k} \geq 1$, $i = 3, 4, \ldots, n-1$, $\forall k \in M$, then the condition (3.3) in Assumption 3.1 can be removed.*

Proof When $i = 1$ and $q_1 = 1$, from Corollary 3.13, the conclusion is correct. For $i = 2$, since $\tau_k - 1 < 0$, one has

$$\frac{r_{2,k}}{q_3} = \tau_k - 1 + \frac{1}{q_2} < \frac{1}{q_2} \leq \frac{1}{q_1} = 1. \tag{3.43}$$

When $i = 3$, similarly we have

$$\frac{r_{3,k}}{q_4} = \tau_k - 1 + \frac{1}{q_3} < \frac{1}{q_3} \leq \frac{1}{q_1} = 1. \tag{3.44}$$

In addition, from (3.2), we obtain $\tau_k = \frac{r_{1,k}}{q_2}$ and then

$$\frac{r_{3,k}}{q_4} = \frac{r_{1,k}}{q_2} - 1 + \frac{1}{q_3} < \frac{r_{1,k}}{q_2} < \frac{1}{q_2}. \tag{3.45}$$

With the help of (3.2), for $i = 2, \ldots, n$, we have

$$\frac{1}{q_i} = \frac{1}{r_{i-1,k}}(\tau_k - 1 + \frac{1}{q_{i-1}}) < \frac{1}{r_{i-1,k}q_{i-1}}. \tag{3.46}$$

When $i \geq 4$, from (3.2) and (3.46), we get

$$\frac{r_{i,k}}{q_{i+1}} = \tau_k - 1 + \frac{1}{q_i} < \frac{1}{q_i} \leq \frac{1}{q_1} = 1, \tag{3.47}$$

and

$$\frac{r_{i,k}}{q_{i+1}} < \frac{1}{q_i} < \frac{1}{r_{i-1,k}q_{i-1}} \leq \frac{1}{q_{i-1}}. \tag{3.48}$$

Note that last inequality in (3.48) holds due to $r_{i-1,k} \geq 1$.

Repeatedly applying (3.46) to (3.48) yields

$$\frac{r_{i,k}}{q_{i+1}} < \frac{1}{r_{l,k}q_l} \leq \frac{1}{q_l}, l = 3, 4, \ldots, i-2. \tag{3.49}$$

From (3.48) together with (3.46), we obtain that

$$\frac{r_{i,k}}{q_{i+1}} < \frac{1}{r_{3,k}}\frac{1}{r_{2,k}q_2} \leq \frac{1}{q_2} \tag{3.50}$$

by $r_{2,k}r_{3,k} \geq 1$. From (3.47), (3.48), (3.49) and (3.50), we conclude that (3.3) can be removed.

Remark 3.18 Corollary 3.17 shows that under certain additional conditions equation (3.2) in Assumption 3.1 is sufficient for switched systems (3.1) where some of the powers are bigger than 1 while others are less than 1. For example, when $m = 2, n = 3$, a switched system (3.1) with $r_{1,1} = \frac{9}{11}, r_{2,1} = \frac{7}{5}, r_{3,1} = 1, r_{1,2} = \frac{29}{33}$, $r_{2,2} = \frac{203}{135}$, and $r_{3,2} = \frac{107}{93}$ satisfies the conditions of Corollary 3.17 and Eq. (3.2) with $q_1 = 1, q_2 = 1, q_3 = \frac{77}{45}, q_4 = \frac{77}{31}, \tau_1 = \frac{9}{11}$ and $\tau_2 = \frac{29}{33}$.

3.5 Extension to Non-smooth Case

In the previous sections, we have developed the finite-time stabilization of (3.1) under arbitrary switchings, where $f_{i,\sigma(t)}(\bar{x}_i)$ are assumed C^1. In this section, we will extend the previous results to the finite-time stabilization of (3.1) containing non-smooth nonlinearities by co-designing stabilizers and a switching policy. We reconsider

$$\dot{x}_i = x_{i+1}^{r_{i,\sigma(t)}} + f_{i,\sigma(t)}(\bar{x}_i), i = 1, \ldots, n, u_{\sigma(t)} = x_{n+1}, \tag{3.51}$$

but where

$$f_{i,\sigma(t)}(\bar{x}_i) = \bar{f}_{i,\sigma(t)}(\bar{x}_i) + \tilde{f}_{i,\sigma(t)}(\bar{x}_i),$$

$\bar{f}_{i,k}$ are C^1 and $\tilde{f}_{i,k}$ are C^0, and vanishing at $x = 0$ for all $k \in M$. For simplification, we rewrite the system (3.51) into a compact form:

$$\dot{x} = \bar{F}_{\sigma(t)}(x, u_{\sigma(t)}) + \tilde{F}_{\sigma(t)}(x), \tag{3.52}$$

where $\bar{F}_{\sigma(t)} = \left[x_2^{r_{1,\sigma(t)}} + \bar{f}_{1,\sigma(t)}, x_3^{r_{2,\sigma(t)}} + \bar{f}_{2,\sigma(t)}, \ldots, u_{\sigma(t)}^{r_{n,\sigma(t)}} + \bar{f}_{n,\sigma(t)}\right]^T$, $\tilde{F}_{\sigma(t)} = \left[\tilde{f}_{1,\sigma(t)}, \tilde{f}_{2,\sigma(t)}, \ldots, \tilde{f}_{n,\sigma(t)}\right]^T$.

The C^1 functions $\bar{f}_{i,k}$ can be viewed as higher order nonlinearities, while the C^0 functions $\tilde{f}_{i,k}$ can be viewed as lower order nonlinearity (such as, $|\tilde{f}_{i,k}| \leq$

$C\sum_{j=1}^{i}|x_j|^{n_{i,j}}$ with $n_{i,j} < 1$). In the special case where the system (3.51) involves no nonlinearities $\tilde{f}_{i,k}$ (i.e., $\tilde{f}_{i,k}(\bar{x}_i) \equiv 0$), such system (3.51) becomes the system (3.1). The structure of the subsystems in system (3.51) is much more general than that of the non-switched systems in [12].

Our control objective is to find under what conditions and how can the co-design of stabilizers and a switching law be proposed to solve the global finite-time stabilization for the system (3.51). If at least one of the subsystems of system (3.51) is finite-time stabilizable based on the existing methods, this problem is trivial. Therefore, none of the individual subsystems is assumed to be finite-time stabilizable.

We first consider the system (3.51) when $\tilde{f}_{i,k}(\bar{x}_i) \equiv 0$, that is, $\dot{x} = \bar{F}_{\sigma(t)}(x, u_{\sigma(t)})$, which essentially is system (3.1). Thus, following the previous design procedure in Sect. 3.3 we will construct the controllers $u_k(x)$ for each subsystem and a common Lyapunov function $V(x)$ such that

$$\dot{V}(x) \leq -c_k V^{(1+\tau_k)/2}(x) \tag{3.53}$$

for finite-time stabilization of switched system $\dot{x} = \bar{F}_{\sigma(t)}(x, u_{\sigma(t)})$, then design a switching law, together with the $u_k(x)$, to finite-time stabilize (3.51) where $\tilde{F}_{\sigma(t)}(x)$ can *destabilize all subsystems of* (3.51) *without switchings*.

Next, we shall prove that under an appropriate condition of a convex combination, global finite-time stabilization for the system (3.51) can be achieved by co-designing stabilizers for the subsystems and a switching law.

Theorem 3.19 *Consider switched system (3.51) satisfying Assumptions 3.1 and 3.2 and suppose that there is a set of scalars ϱ_k, $k = 1, 2, \ldots, m$, such that*

$$\varrho_1 + \varrho_2 + \cdots + \varrho_{m-1} + \varrho_m = 1, \tag{3.54}$$

$$\frac{\partial V}{\partial x}\left[\varrho_1 \tilde{F}_1(x) + \varrho_2 \tilde{F}_2(x) + \cdots + \varrho_m \tilde{F}_m(x)\right] \leq 0. \tag{3.55}$$

Then, the state feedback controllers $u_k(x)$ for the subsystems and a switching law

$$\sigma(t) = arg\min_{k\in M}\left\{\frac{\partial V}{\partial x}\tilde{F}_k(x)\right\} \tag{3.56}$$

guarantee that the closed-loop system (3.51) is globally finite-time stabilized.

Proof Based on the constructed Lyapunov function $V(x)$ and the controllers $u_k(x)$ for subsystems, when $\sigma(t) = k$, one has

$$\dot{V}(x) = \frac{\partial V}{\partial x}\bar{F}_k(x, u_k(x)) + \frac{\partial V}{\partial x}\tilde{F}_k(x). \tag{3.57}$$

Since $0 < \varrho_k < 1$, it follows that for each $x \in R^n$, at least one of the quantities $\frac{\partial V}{\partial x}\tilde{F}_k(x)$ are non-positive. In other words, R^n is covered by the union of closed regions $\Omega_k = \{x|\frac{\partial V}{\partial x}\tilde{F}_k(x) \le 0\}, k = 1, 2, \ldots, m$. Then, the function $V(x)$ decreases along solutions of the closed-loop subsystem k in the region Ω_k and

$$\dot{V} \le -c_k V^{(1+\tau_k)/2}, \tag{3.58}$$

which implies that the closed-loop switched system (3.51) under the switching law is global finite-time stable.

3.6 Illustrative Examples

In this section, two examples are studied to show the effectiveness of the proposed methods. One is to show how to construct a common Lyapunov function and a finite-time stabilizer to achieve global finite-time stabilization under arbitrary switchings; the other is to show how to co-design finite-time stabilizers and a switching signal to guarantee global finite-time stabilization.

Example 3.20 *(Finite-time stabilization under arbitrary switchings)* Consider the following switched system:

$$\begin{aligned} \dot{x}_1 &= x_2^{r_{1,\sigma(t)}} + f_{1,\sigma(t)}(x_1), \\ \dot{x}_2 &= x_3^{r_{2,\sigma(t)}} + f_{2,\sigma(t)}(x_1, x_2), \\ \dot{x}_3 &= u_{\sigma(t)}^{r_{3,\sigma(t)}} + f_{3,\sigma(t)}(x_1, x_2, x_3), \end{aligned} \tag{3.59}$$

where $\sigma(t) : [0, \infty) \to M = \{1, 2\}$, $f_{1,1}(x_1) = x_1^3$, $f_{1,2}(x_1) = x_1 \sin x_1$, $f_{2,1}(x_1, x_2) = x_1 \sin(x_2)$, $f_{2,2}(x_1, x_2) = x_1 \cos(x_1 x_2)$, $f_{3,1} = f_{3,2} = 0, r_{1,1} = r_{2,1} = r_{3,1} = \frac{1}{3}$, and $r_{1,2} = r_{2,2} = r_{3,2} = \frac{1}{5}$.

We will design a state feedback control law to globally finite-time stabilize the switched system (3.59) under arbitrary switchings.

It is easy to see that the switched system (3.59) cannot be finite-time stabilized via the existing methods in [15, 16]. However, it is still possible to design the state-feedback controllers for subsystems to reach global finite-time stability under arbitrary switchings.

One can verify that Assumption 3.1 holds with $q_1 = q_2 = q_3 = q_4 = 1$, $\tau_1 = \frac{1}{3}$ and $\tau_2 = \frac{1}{5}$.

Following the design procedure in Sect. 3.3, the controller for system (3.59) can be constructed as

$$u_k = x_3^* = -\beta_3(x)\xi_3^{\frac{1}{q_4}}, k = 1, 2, \tag{3.60}$$

where $\xi_2 = x_2 + \beta_1(x_1)\xi_1$, $\xi_3 = x_3 + \beta_2(x_1, x_2)\xi_2$ with $\beta_1(x_1) = (7 + 2x_1^2)^3 \left[1 + (5 + x_1^2)^2\right]$, $\beta_2(x_1, x_2) = \left[150 + \lambda_2\left((3/4)^{4/3}(2/3 + 1/3\lambda_2) + 1\right) + 2\tilde{c}_2 + \bar{c}_2 + 2\right]^5$, $\lambda_2 = \beta_1(x_1) + \left[12x_1^2(7 + 2x_1^2)^2(1 + (5 + x_1^2)^2) + 4x_1^2(7 + 2x_1^2)^3(5 + x_1^2)\right]\beta_1^3$

(x_1), $\bar{c}_2 = 2/5 + 2/3\big[(34/15 + 1/3x_1^2 + 2/5\xi_2^2)(1 + x_1^2)\beta_1(x_1)\big]^2$, $\beta_3(x) = (150 + 3.5(\lambda_3^{1+\tau_1} + \lambda_4^{1+\tau_1}) + 1)^5$, $\lambda_3 = (\beta_1(x_1)\beta_2(x_1, x_2) + 12x_1^2(7 + 2x_1^2)^2(1 + (5 + x_1^2)^2)\beta_2(x_1, x_2) + 4x_1^2(7 + 2x_1^2)^3(5 + x_1^2) + \frac{1}{2}(\frac{\partial \beta_2(\bar{x}_2)}{\partial x_1})^2 + \frac{1}{2}\xi_2^2)\beta_1(x_1)$, and $\lambda_4 = \beta_2^2 + 5\beta_2(150 + \lambda_2((3/4)^{4/3}(2 + \lambda_2)/3 + 1) + 2\tilde{c}_2 + \bar{c}_2 + 2)^4\big[56/15(34/15 + x_1^2/3 + 2/5\xi_2^2)(1 + x_1^2)^2\beta_1^2\xi_2^2 + (1 + \xi_2^2)/5\big]$, such that the closed-loop switched system (3.59) with (3.60) satisfies

$$\dot{V}_3(x) \leq -c_k V_3^{\frac{1+\tau_k}{2}}(x), \tag{3.61}$$

where

$$V_3 = \frac{1}{2}\left(x_1^2 + \xi_2^2 + \xi_3^2\right), \tag{3.62}$$

and $c_k = 2^{-\frac{1+\tau_k}{2}}$. According to Theorem 3.10, the closed-loop switched system (3.59) with (3.60) is globally finite-time stable under arbitrary switchings.

The simulation is carried out with the initial state $x_0 = (-0.3, 0.3, 0.5)^T$. Figure 3.1 shows the state trajectories of the closed-loop switched system (3.59) with (3.60) under a randomly chosen switching signal of time, shown in Fig. 3.2. It can be clearly observed from Fig. 3.1 that the finite stabilization has been achieved. Thus, the simulation results well illustrate the effectiveness of the proposed method.

Example 3.21 *(Finite-time stabilization by co-designing stabilizers and a switching signal)* Consider the following switched nonlinear system:

$$\begin{aligned} \dot{x}_1 &= x_2^{r_{1,\sigma(t)}} + f_{1,\sigma(t)}(x_1), \\ \dot{x}_2 &= u_{\sigma(t)}^{r_{2,\sigma(t)}}, \end{aligned} \tag{3.63}$$

where $\sigma(t) : [0, \infty) \to \{1, 2\}$, $r_{1,1} = r_{2,1} = 1/3$, $r_{1,2} = r_{2,2} = 3/5$, $f_{1,k} = \bar{f}_{1,k} + \tilde{f}_{1,k}$ with $\bar{f}_{1,1} = \bar{f}_{1,2} = 0$, $\tilde{f}_{1,1}(x_1) = -7|x_1|^{3/5}$, $\tilde{f}_{1,2}(x_1) = 21|x_1|^{3/5}$.

Because of the existence of $\tilde{f}_{1,1}$ and $\tilde{f}_{1,2}$ which are only C^0, but not C^1, the system (3.63) cannot be handled with the results obtained above and those methods in [12, 13, 22]. Therefore, an interesting and useful idea is to solve the global finite-time stabilization problem by co-designing stabilizers for subsystems and a switching law.

It is easy to verify that Assumption 3.1 holds with $q_1 = q_2 = q_3 = 1$, $\tau_1 = 1/3$, $\tau_2 = 3/5$. We first design controllers for each subsystem of the switched system:

$$\begin{aligned} \dot{x}_1 &= x_2^{r_{1,\sigma(t)}}, \\ \dot{x}_2 &= u_{\sigma(t)}^{r_{2,\sigma(t)}}. \end{aligned} \tag{3.64}$$

According to the design procedure above, we obtain that

$$V = \frac{1}{2}x_1^2 + \frac{1}{2}(x_2 + 8x_1)^2,$$

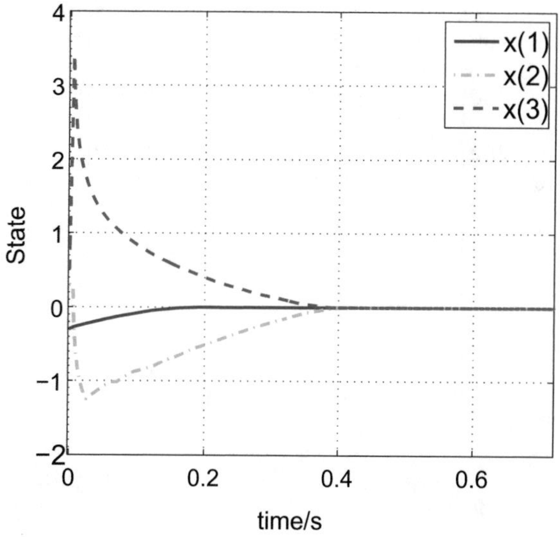

Fig. 3.1 The trajectories of the state x

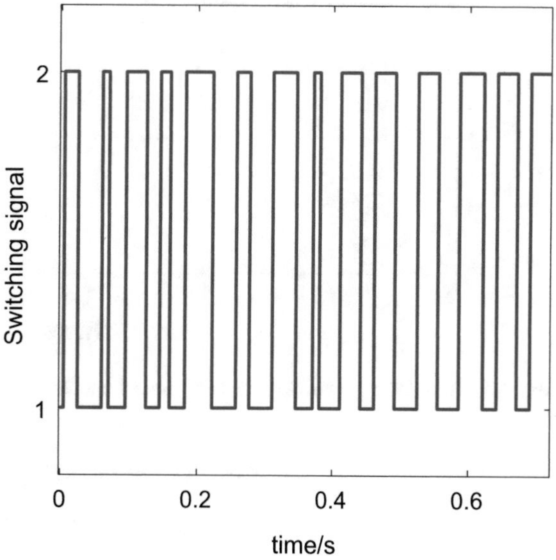

Fig. 3.2 The switching signal for the switched system

$x_2^* = -8x_1$ and the controller

$$u_k = -\beta_2(x)\xi_2 = -\beta_2(x)(x_2 + 8x_1), k = 1, 2, \tag{3.65}$$

where $\beta_2(x) \geq \max_{k \in M} \left(c_{2,1,k} + c_{2,2,k} + 1\right)^{\frac{1}{r_{2,k}}}$ with $c_{2,1,1} = 27/5$, $c_{2,1,2} = 23/10$, $c_{2,2,1} = 27.2$ and $c_{2,2,2} = 41.4$, such that the closed-loop system (3.64) satisfies

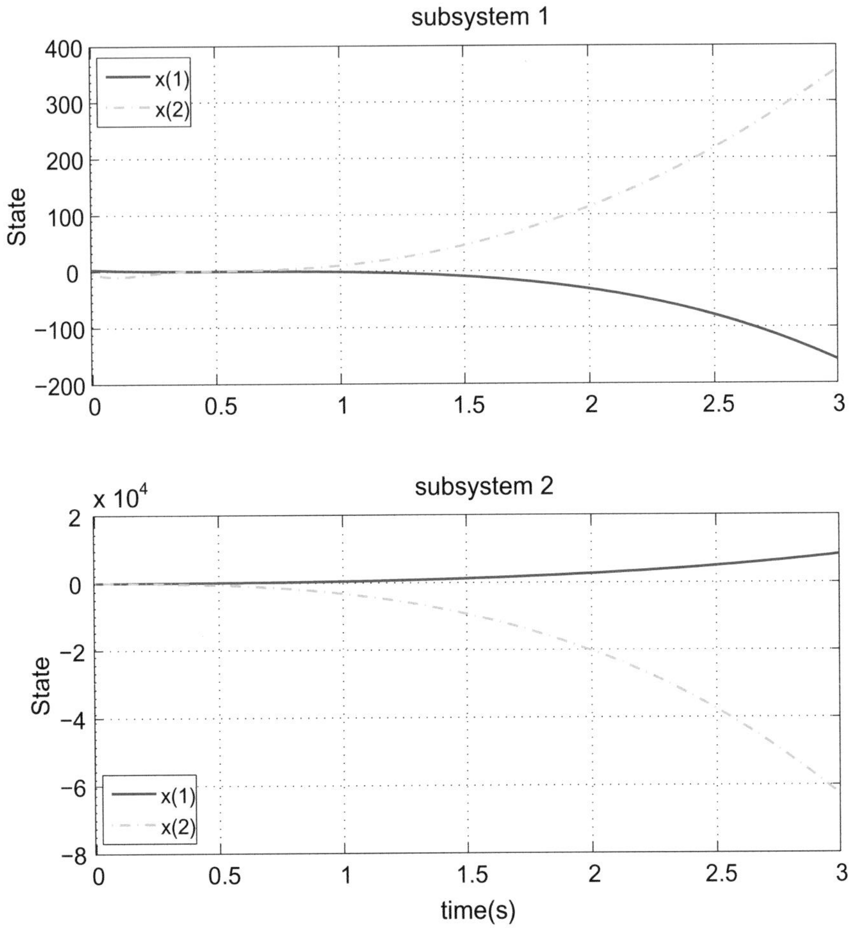

Fig. 3.3 The state trajectories of the subsystem 1 and 2

$$\dot{V}|_{\sigma=1} \leq -2^{-2/3}V^{2/3}(x),$$
$$\dot{V}|_{\sigma=2} \leq -2^{-4/5}V^{4/5}(x). \tag{3.66}$$

On the other hand, one can verify that $\frac{\partial V}{\partial x}\left[\varrho_1 \tilde{F}_1(x) + \varrho_2 \tilde{F}_2(x)\right] = 0$ with $\tilde{F}_1 = \left[\tilde{f}_{1,1}, 0\right]^T$, $\tilde{F}_2 = \left[\tilde{f}_{1,2}, 0\right]^T$, $\varrho_1 = \frac{3}{4}$ and $\varrho_2 = \frac{1}{4}$. Therefore, according to Theorem 3.19, the closed-loop switched system (3.63) under the switching law (3.56) is global finite-time stable. Figures 3.3 and 3.4 depict the simulation results with the initial state $x_0 = (2.6, -5)^T$. From Fig. 3.3, it is easy to see that each closed-loop subsystem is unstable. From Fig. 3.4, we can easily to see that the closed-loop switched system (3.63) with unstable subsystems is finite-time stabilized under the stabilizers (3.65) and the switching law (3.56).

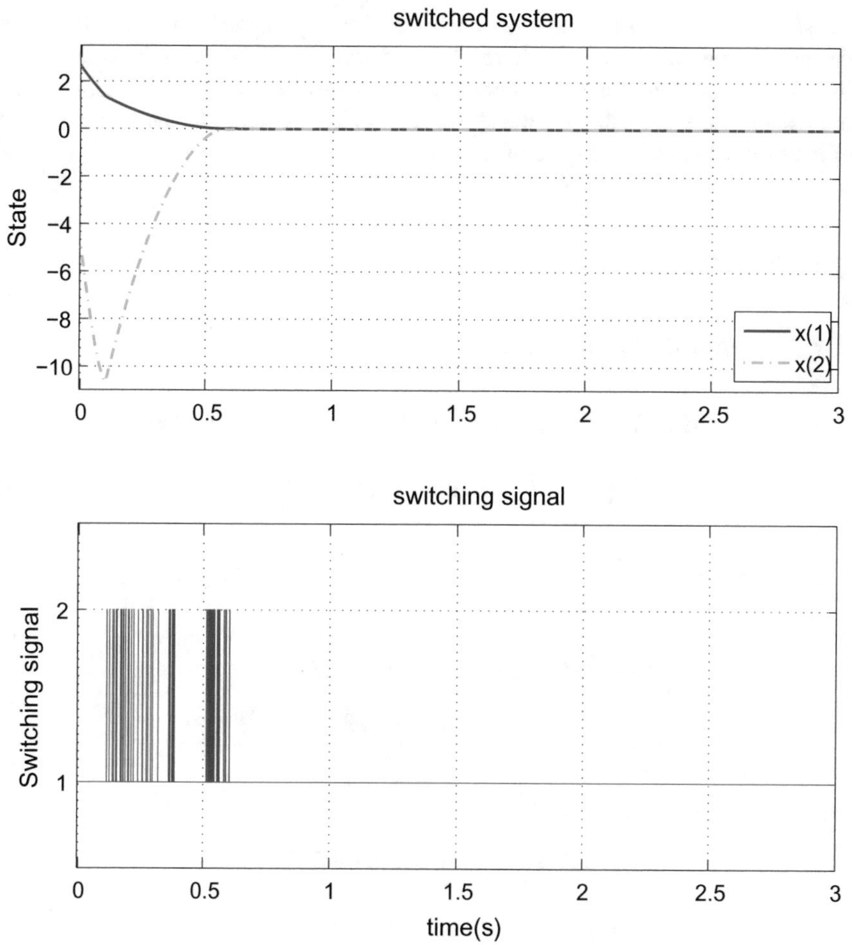

Fig. 3.4 The state trajectories and the switching signal of the switched system

3.7 Concluding Remarks

This chapter has studied the problem of global finite-time stabilization of a class of switched strict-feedback nonlinear systems, whose subsystems have chained integrators with the powers of positive odd rational numbers and where all the powers in $\dot{x}_i$-equations of subsystems can be different. Based on the common Lyapunov function method and the technique of adding a power integrator, we have exploited a recursive design approach for designing the global finite-time stabilizer, which guarantees the global finite-time stability of the corresponding closed-loop switched system. Construction of a common coordinate transformation for all subsystems plays a crucial role in the method, which avoids dealing with different coordinate transformations for different subsystems required by application of the recursive design scheme. In addition, we have proposed some conditions that, by characterizing the powers of the chained integrators of the considered switched system, enable our design. Last, we extended the above-mentioned results to the non-smooth case of the switched nonlinear systems.

3.8 Proof of Propositions

3.8.1 Proof of Proposition 3.5

Proof Recall that $W_i(\bar{x}_i)$ in (3.12), it is straightforward to show that

$$\frac{\partial W_i(\bar{x}_i)}{\partial x_i} = \left(x_i^{q_i} - x_i^{*q_i}\right)^{2-\frac{1}{q_i}} = \xi_i^{2-\frac{1}{q_i}}. \tag{3.67}$$

For simplification, we denote $x_i^*(\Delta_j) = x_i^*(x_1, \ldots, x_{j-1},\ x_j + \Delta, x_{j+1}, \ldots, x_{i-1})$. We calculate the limit

$$\begin{aligned} &\lim_{\Delta\to 0} \frac{W_i(x_1,\ldots,x_{j-1},x_j+\Delta,x_{j+1},\ldots,x_i)-W_i(\bar{x}_i)}{\Delta} \\ &= \lim_{\Delta\to 0} \frac{\int_{x_i^*(\Delta_j)}^{x_i} \left(s^{q_i}-x_i^{*q_i}(\Delta_j)\right)^{2-\frac{1}{q_i}} ds - \int_{x_i^*}^{x_i} \left(s^{q_i}-x_i^{*q_i}\right)^{2-\frac{1}{q_i}} ds}{\Delta} \\ &= \lim_{\Delta\to 0} \frac{\int_{x_i^*(\Delta_j)}^{x_i^*} \left(s^{q_i}-x_i^{*q_i}(\Delta_j)\right)^{2-\frac{1}{q_i}} ds}{\Delta} \\ &\quad + \lim_{\Delta\to 0} \frac{\int_{x_i^*}^{x_i} \left(s^{q_i}-x_i^{*q_i}(\Delta_j)\right)^{2-\frac{1}{q_i}} - \left(s^{q_i}-x_i^{*q_i}\right)^{2-\frac{1}{q_i}} ds}{\Delta} \\ &= \lim_{\Delta\to 0} \frac{\int_{x_i^*(\Delta_j)}^{x_i^*} \left(s^{q_i}-x_i^{*q_i}(\Delta_j)\right)^{2-\frac{1}{q_i}} ds}{\Delta} \\ &\quad - \left(2-\frac{1}{q_i}\right) \frac{\partial(x_i^{*q_i})}{\partial x_j} \int_{x_i^*}^{x_i} \left(s^{q_i} - x_i^{*q_i}\right)^{1-\frac{1}{q_i}} ds. \end{aligned} \tag{3.68}$$

It is easy to obtain that

$$\begin{aligned} \frac{\int_{x_i^*(\Delta_j)}^{x_i^*} \left(s^{q_i}-x_i^{*q_i}(\Delta_j)\right)^{2-1/q_i} ds}{\Delta} &\le \left|\frac{x_i^{*q_i}-x_i^{*q_i}(\Delta_j)}{\Delta}\right| \\ &\quad \times \left|x_i^{*q_i} - x_i^{*q_i}(\Delta_j)\right|^{1-\frac{1}{q_i}} \left|x_i^* - x_i^*(\Delta_j)\right|. \end{aligned} \tag{3.69}$$

Note again that $x_i^{*q_i}$ is C^1. Hence, from (3.69), one has

$$\lim_{\Delta\to 0} \frac{\int_{x_i^*(\Delta_j)}^{x_i^*} \left(s^{q_i}-x_i^{*q_i}(\Delta_j)\right)^{2-1/q_i} ds}{\Delta} = 0. \tag{3.70}$$

So, from (3.68), we obtain that

$$\frac{\partial W_i}{\partial x_j} = -\left(2 - \frac{1}{q_i}\right) \frac{\partial (x_i^{*q_i})}{\partial x_l} \int_{x_i^*}^{x_i} \left(s^{q_i} - x_i^{*q_i}\right)^{1-\frac{1}{q_i}} ds. \tag{3.71}$$

As a consequence, $W_i(\bar{x}_i)$ is C^1 because of continuity of $\frac{\partial W_i(\bar{x}_i)}{\partial x_j}$, $j = 1, 2, \ldots, i-1$. So $V_i(\bar{x}_i)$ is also C^1.

Next, we prove that $V_i(\bar{x}_i)$ is positive definite and proper.

When $x_i^* \le x_i$, using (1.8) leads to

$$\begin{aligned} W_i(\bar{x}_i) &= \int_{x_i^*}^{x_i} \left(s^{q_i} - x_i^{*q_i}\right)^{2-\frac{1}{q_i}} ds \\ &\ge \left(2^{1-q_i}\right)^{2-\frac{1}{q_i}} \int_{x_i^*}^{x_i} \left(s - x_i^*\right)^{2q_i - 1} ds \\ &= 2^{3-\frac{1}{q_i}-2q_i} \int_{x_i^*}^{x_i} \left(s - x_i^*\right)^{2q_i-1} ds \\ &= 2^{3-\frac{1}{q_i}-2q_i} \frac{1}{2q_i} \left(x_i - x_i^*\right)^{2q_i}. \end{aligned} \tag{3.72}$$

On the other hand, it can be shown that (3.72) holds as well when $x_i^* > x_i$, that is,

$$\begin{aligned} W_i(\bar{x}_i) &= \int_{x_i^*}^{x_i} \left(s^{q_i} - x_i^{*q_i}\right)^{2-\frac{1}{q_i}} ds \\ &= \int_{x_i}^{x_i^*} \left(x_i^{*q_i} - s^{q_i}\right)^{2-\frac{1}{q_i}} ds \\ &\ge \left(2^{1-q_i}\right)^{2-\frac{1}{q_i}} \int_{x_i}^{x_i^*} \left(x_i^* - s\right)^{2q_i-1} ds \\ &= -2^{3-\frac{1}{q_i}-2q_i} \int_{x_i}^{x_i^*} \left(x_i^* - s\right)^{2q_i-1} d\left(x_i^* - s\right) \\ &= -2^{3-\frac{1}{q_i}-2q_i} \frac{1}{2q_i} \left[0 - \left(x_i^* - x_i\right)^{2q_i}\right] \\ &= 2^{3-\frac{1}{q_i}-2q_i} \frac{1}{2q_i} \left(x_i - x_i^*\right)^{2q_i}. \end{aligned} \tag{3.73}$$

Therefore, $V_i(\bar{x}_i) \ge V_{i-1}(\bar{x}_{i-1}) + m_i \left(x_i - x_i^*\right)^{2q_i}$ with a positive constant m_i, which implies that $V_i(\bar{x}_i)$ is positive definite and proper. The proof of proposition is completed.

3.8.2 *Proof of Proposition 3.6*

Proof By using Assumption 3.1 and (1.8), we have

$$\begin{aligned} &\left|x_i^{r_{i-1,k}} + (-x_i^*)^{r_{i-1,k}}\right| \\ &= \left|\left(x_i^{q_i}\right)^{\frac{r_{i-1,k}}{q_i}} + \left((-x_i^*)^{q_i}\right)^{\frac{r_{i-1,k}}{q_i}}\right| \\ &\le 2^{1-\frac{r_{i-1,k}}{q_i}} \left|x_i^{q_i} + (-x_i^*)^{q_i}\right|^{\frac{r_{i-1,k}}{q_i}} \\ &= 2^{1-\frac{r_{i-1,k}}{q_i}} \left|x_i^{q_i} - x_i^{*q_i}\right|^{\frac{r_{i-1,k}}{q_i}} \end{aligned}$$

$$= 2^{1-\frac{r_{i-1,k}}{q_i}} |\xi_i|^{\frac{r_{i-1,k}}{q_i}}. \tag{3.74}$$

Thus, from (1.9) and (3.74), we can obtain that

$$\begin{aligned}
&\left|\xi_{i-1}^{2-\frac{1}{q_{i-1}}}(x_i^{r_{i-1,k}} - x_i^{*r_{i-1,k}})\right| \\
&\leq 2^{1-\frac{r_{i-1,k}}{q_i}} |\xi_{i-1}|^{2-\frac{1}{q_{i-1}}} |\xi_i|^{\frac{r_{i-1,k}}{q_i}} \\
&\leq 2^{1-\frac{r_{i-1,k}}{q_i}} \Bigg(\tfrac{2-1/q_{i-1}}{1+\tau_k}\omega_k |\xi_{i-1}|^{1+\tau_k} + \tfrac{r_{i-1,k}/q_i}{1+\tau_k} \\
&\quad \times \omega_k^{-\frac{2-1/q_{i-1}}{r_{i-1,k}/q_i}} |\xi_i|^{1+\tau_k} \Bigg) \\
&\leq \tfrac{1}{3} |\xi_{i-1}|^{1+\tau_k} + c_{i1,k} |\xi_i|^{1+\tau_k} \\
&\leq \tfrac{1}{3} \left(\xi_1^{1+\tau_k} + \cdots + \xi_{i-1}^{1+\tau_k} \right) + c_{i1,k} \xi_i^{1+\tau_k},
\end{aligned} \tag{3.75}$$

where $\omega_k = \frac{1}{3}\frac{1+\tau_k}{2-1/q_{i-1}} 2^{\frac{r_{i-1,k}}{q_i}-1}$, $c_{i1,k} = 2^{1-\frac{r_{i-1,k}}{q_i}} \frac{r_{i-1,k}/q_i}{1+\tau_k} \omega_k^{-\frac{2-1/q_{i-1}}{r_{i-1,k}/q_i}}$, $k = 1, 2, \ldots, m$.

3.8.3 *Proof of Proposition 3.7*

Proof From (3.15) and $x_1^* = 0$, we conclude that

$$\begin{aligned}
\left|\xi_i^{2-\frac{1}{q_i}} f_{i,k}\right| &\leq |\xi_i|^{2-\frac{1}{q_i}} (|x_1| + \cdots + |x_i|)\gamma_{i,k}(\bar{x}_i) \\
&\leq |\xi_i|^{2-\frac{1}{q_i}} \left[\sum_{j=1}^{i} |x_j - x_j^*| + \sum_{j=2}^{i} |x_j^*| \right] \gamma_{i,k}(\bar{x}_i),
\end{aligned} \tag{3.76}$$

where the first inequality follows from (3.5) and the last inequality follows from the fact that $|a + b| \leq |a| + |b|$.

Using (1.8) with $x = x_i$, $y = x_j^*$, $q = q_j$, and $x_i^* = -\beta_{i-1}(\bar{x}_{i-1})\xi_{i-1}^{\frac{1}{q_i}}$, we then obtain from (3.76):

$$\begin{aligned}
&\left|\xi_i^{2-\frac{1}{q_i}} f_{i,k}\right| \\
&\leq |\xi_i|^{2-\frac{1}{q_i}} \left[\sum_{j=1}^{i} 2^{\frac{q_j-1}{q_j}} |\xi_j|^{\frac{1}{q_j}} + \sum_{j=2}^{i} |\xi_{j-1}|^{\frac{1}{q_j}} \beta_{j-1} \right] \gamma_{i,k}.
\end{aligned} \tag{3.77}$$

By using the fact that $0 < \frac{r_{i,k}}{q_{i+1}} < 1$, $i = 1, 2, \ldots, n$, in (3.4) and $\frac{r_{i,k}}{q_{i+1}} \leq \min \left\{\frac{1}{q_1}, \frac{1}{q_2}, \ldots, \frac{1}{q_i}\right\} \leq 1, \forall i = 1, \ldots, n, \forall k \in M$ in (3.3), we apply (1.10) to each term in the summation of (3.77) and obtain

$$\begin{aligned} |\xi_j|^{\frac{1}{q_j}} &\leq |\xi_j|^{\frac{r_{i,k}}{q_{i+1}}} + |\xi_j| = |\xi_j|^{\frac{r_{i,k}}{q_{i+1}}} (1 + |\xi_j|^{1-\frac{r_{i,k}}{q_{i+1}}}), \\ & j = 1, 2, \ldots, i, \\ |\xi_{j-1}|^{\frac{1}{q_j}} &\leq |\xi_{j-1}|^{\frac{r_{i,k}}{q_{i+1}}} + |\xi_{j-1}| \\ &= |\xi_{j-1}|^{\frac{r_{i,k}}{q_{i+1}}} (1 + |\xi_{j-1}|^{1-\frac{r_{i,k}}{q_{i+1}}}), \ j = 2, \ldots, i. \end{aligned} \tag{3.78}$$

Substituting (3.78) into (3.77) yields

$$\begin{aligned} &\left|\xi_i^{2-\frac{1}{q_i}} f_{i,k}\right| \\ &\leq \left[\sum_{j=1}^{i} 2^{\frac{q_j-1}{q_j}} |\xi_j|^{\frac{r_{i,k}}{q_{i+1}}} (1 + |\xi_j|^{1-\frac{r_{i,k}}{q_{i+1}}}) \right. \\ &\quad \left. + \sum_{j=2}^{i} |\xi_{j-1}|^{\frac{r_{i,k}}{q_{i+1}}} (1 + |\xi_{j-1}|^{1-\frac{r_{i,k}}{q_{i+1}}})\beta_{j-1}(\bar{x}_{j-1})\right] |\xi_i|^{2-\frac{1}{q_i}} \gamma_{i,k}(\bar{x}_i). \end{aligned} \tag{3.79}$$

There exist C^1 functions $\tilde{\gamma}_{i,k}(\bar{x}_i) \geq 0$, such that

$$\begin{aligned} \tilde{\gamma}_{i,k}(\bar{x}_i) &\geq 2^{\frac{q_j-1}{q_j}} (1 + |\xi_j|^{1-\frac{r_{i,k}}{q_{i+1}}})\gamma_{i,k}(\bar{x}_i), \\ \tilde{\gamma}_{i,k}(\bar{x}_i) &\geq (1 + |\xi_{j-1}|^{1-\frac{r_{i,k}}{q_{i+1}}})\beta_{j-1}(\bar{x}_{j-1})\gamma_{i,k}(\bar{x}_i) \end{aligned} \tag{3.80}$$

for $j = 1, \ldots, i-1$ (with $\beta_0 = 0$ for convenience). One way to choose a C^1 $\tilde{\gamma}_{i,k}(\bar{x}_i) = 4 \times \left[1 + \frac{1}{2}(1 + \frac{r_{i,k}}{q_{i+1}}) \times 1^2 + \frac{1}{2}(1 - \frac{r_{i,k}}{q_{i+1}})\xi_j^2 + \left(1 + \frac{1}{2}(1 + \frac{r_{i,k}}{q_{i+1}})1^2 + \frac{1}{2}(1 - \frac{r_{i,k}}{q_{i+1}})\xi_{j-1}^2\right) \times \beta_{j-1}\right]\gamma_{i,k}(\bar{x}_i)$ since

$$\begin{aligned} |\xi_j|^{1-\frac{r_{i,k}}{q_{i+1}}} &= 1^{1+\frac{r_{i,k}}{q_{i+1}}} |\xi_j|^{1-\frac{r_{i,k}}{q_{i+1}}} \\ &\leq \frac{1}{2}(1 + \frac{r_{i,k}}{q_{i+1}}) \times 1^2 + \frac{1}{2}(1 - \frac{r_{i,k}}{q_{i+1}})\xi_j^2, \end{aligned} \tag{3.81}$$

and

$$\begin{aligned} |\xi_{j-1}|^{1-\frac{r_{i,k}}{q_{i+1}}} &= 1^{1+\frac{r_{i,k}}{q_{i+1}}} |\xi_{j-1}|^{1-\frac{r_{i,k}}{q_{i+1}}} \\ &\leq \frac{1}{2}(1 + \frac{r_{i,k}}{q_{i+1}}) \times 1^2 + \frac{1}{2}(1 - \frac{r_{i,k}}{q_{i+1}})\xi_{j-1}^2 \end{aligned} \tag{3.82}$$

and $2^{\frac{q_j-1}{q_j}} \leq 2$. Then, we have

$$
\begin{aligned}
|\xi_i|^{2-\frac{1}{q_i}} |\xi_j|^{\frac{r_{i,k}}{q_{i+1}}} \tilde{\gamma}_{i,k} &\leq \frac{1}{6}\xi_j^{1+\tau_k} + \xi_i^{1+\tau_k}\tilde{c}_{i,2,k}(\bar{x}_i), \\
|\xi_i|^{2-\frac{1}{q_i}} |\xi_{j-1}|^{\frac{r_{i,k}}{q_{i+1}}} \tilde{\gamma}_{i,k} &\leq \frac{1}{6}\xi_{j-1}^{1+\tau_k} + \xi_i^{1+\tau_k}\bar{c}_{i,2,k}(\bar{x}_i),
\end{aligned} \tag{3.83}
$$

for C^1 functions $\tilde{c}_{i,2,k}(\bar{x}_i), \bar{c}_{i,2,k}(\bar{x}_i) \geq 0$. We then obtain

$$
\begin{aligned}
&\left|\xi_i^{2-\frac{1}{q_i}} f_{i,k}(\bar{x}_i)\right| \\
&\leq \sum_{j=1}^{i} \xi_i^{2-\frac{1}{q_i}} |\xi_j|^{\frac{r_{i,k}}{q_{i+1}}} \tilde{\gamma}_{i,k} + \sum_{j=2}^{i} \xi_i^{2-\frac{1}{q_i}} |\xi_{j-1}|^{\frac{r_{i,k}}{q_{i+1}}} \tilde{\gamma}_{i,k} \\
&\leq \sum_{j=1}^{i} \left[\frac{1}{6}\xi_j^{1+\tau_k} + \xi_i^{1+\tau_k}\tilde{c}_{i,2,k}\right] + \sum_{j=2}^{i} \left[\frac{1}{6}\xi_{j-1}^{1+\tau_k} + \xi_i^{1+\tau_k}\bar{c}_{i,2,k}\right] \\
&\leq \frac{1}{3}\sum_{j=1}^{i-1} \xi_j^{1+\tau_k} + \xi_i^{1+\tau_k} c_{i,2,k}(\bar{x}_i),
\end{aligned} \tag{3.84}
$$

where $c_{i,2,k}(\bar{x}_i) > 0$ is an appropriate C^1 function.

3.8.4 *Proof of Proposition 3.8*

Proof From (3.12), it follows that

$$
\begin{aligned}
&\left|\sum_{j=1}^{i-1} \frac{\partial W_i}{\partial x_j}\dot{x}_j\right| \\
&\leq \sum_{j=1}^{i-1} \left|\int_{x_i^*}^{x_i} (2-1/q_i)(s^{q_i} - x_i^{*q_i})^{1-\frac{1}{q_i}} ds\right| \\
&\quad \times \left|\frac{\partial(x_i^{*q_i})}{\partial x_j}\right| \left|x_{j+1}^{r_{j,k}} + f_{j,k}\right|.
\end{aligned} \tag{3.85}
$$

The objective is to find an upper bound for each term in (3.85).

At first, note that

$$
\begin{aligned}
&\left|\int_{x_i^*}^{x_i} (2-1/q_i)(s^{q_i}-x_i^{*q_i})^{1-\frac{1}{q_i}}ds\right| \\
&\le (2-1/q_i)\left|x_i-x_i^*\right|\left|x_i^{q_i}-x_i^{*q_i}\right|^{1-\frac{1}{q_i}} \\
&= (2-1/q_i)\left|(x_i^{q_i})^{\frac{1}{q_i}}-(x_i^{*q_i})^{\frac{1}{q_i}}\right|\left|x_i^{q_i}-x_i^{*q_i}\right|^{1-\frac{1}{q_i}} \\
&\le (2-1/q_i)2^{1-\frac{1}{q_i}}\left|x_i^{q_i}-x_i^{*q_i}\right|^{\frac{1}{q_i}}\left|x_i^{q_i}-x_i^{*q_i}\right|^{1-\frac{1}{q_i}} \\
&= (2-1/q_i)2^{1-\frac{1}{q_i}}\left|\xi_i\right|.
\end{aligned}
\tag{3.86}
$$

Note that the first inequality and the second one of (3.86) are derived with the help of the first mean value theorem of integration and (1.8), respectively.

We observe that $1-\frac{1}{q_i}\ge 0$. By the inductive assumption in (3.10), we compute the bound of $\left|\frac{\partial(x_i^{*q_i})}{\partial x_j}\right|$ as follows: For $j=1,2,\ldots,i-2$:

$$
\begin{aligned}
&\left|\frac{\partial(x_i^{*q_i})}{\partial x_j}\right| = \left|\frac{\partial}{\partial x_j}(\beta_{i-1}^{q_i}(\bar{x}_{i-1})\xi_{i-1})\right| \\
\le &\left|\frac{\partial \xi_{i-1}}{\partial x_j}\right|\beta_{i-1}^{q_i}(\bar{x}_{i-1})+\left|\frac{\partial\beta_{i-1}^{q_i}(\bar{x}_{i-1})}{\partial x_j}\right|\left|\xi_{i-1}\right| \\
= &\left|\frac{\partial x_{i-1}^{*q_{i-1}}}{\partial x_j}\right|\beta_{i-1}^{q_i}(\bar{x}_{i-1})+\left|\frac{\partial\beta_{i-1}^{q_i}(\bar{x}_{i-1})}{\partial x_j}\right|\left|\xi_{i-1}\right| \\
\le &\left[\left|\frac{\partial x_{i-2}^{*q_{i-2}}}{\partial x_j}\right|\beta_{i-2}^{q_{i-1}}(\bar{x}_{i-2})+\left|\frac{\partial\beta_{i-2}^{q_{i-1}}(\bar{x}_{i-2})}{\partial x_j}\right|\left|\xi_{i-2}\right|\right] \\
&\times\beta_{i-1}^{q_i}(\bar{x}_{i-1})+\left|\frac{\partial\beta_{i-1}^{q_i}(\bar{x}_{i-1})}{\partial x_j}\right|\left|\xi_{i-1}\right| \\
\le &\left|\frac{\partial x_{i-2}^{*q_{i-2}}}{\partial x_j}\right|\beta_{i-2}^{q_{i-1}}(\bar{x}_{i-2})\beta_{i-1}^{q_i}(\bar{x}_{i-1})+\left|\frac{\partial\beta_{i-2}^{q_{i-1}}(\bar{x}_{i-2})}{\partial x_j}\right| \\
&\times\left|\xi_{i-2}\right|\beta_{i-1}^{q_i}(\bar{x}_{i-1})+\left|\frac{\partial\beta_{i-1}^{q_i}(\bar{x}_{i-1})}{\partial x_j}\right|\left|\xi_{i-1}\right| \\
\le &B_j(\bar{x}_j)\left|x_j\right|^{q_j-1}+\sum_{l=0}^{i-j-1}\left|\xi_{j+l}\right|\left|\frac{\partial\beta_{j+l}^{q_{j+l+1}}(\bar{x}_{j+l})}{\partial x_j}\right|\tilde{\alpha}_{j+l+1},
\end{aligned}
\tag{3.87}
$$

where $B_j(\bar{x}_j)$ and $\tilde{\alpha}_j(\bar{x}_j)$ are positive smooth functions.

For $j=i-1$:

$$\begin{aligned}
\left|\frac{\partial(x_i^{*q_i})}{\partial x_{i-1}}\right| &= \left|\frac{\partial}{\partial x_{i-1}}(\beta_{i-1}^{q_i}(\bar{x}_{i-1})\xi_{i-1})\right| \\
&\leq \left|\frac{\partial \xi_{i-1}}{\partial x_{i-1}}\right| \beta_{i-1}^{q_i}(\bar{x}_{i-1}) + \left|\frac{\partial \beta_{i-1}^{q_i}(\bar{x}_{i-1})}{\partial x_{i-1}}\right| |\xi_{i-1}| \\
&\leq \left|\frac{\partial(x_{i-1}^{q_{i-1}} - x_{i-1}^{*q_{i-1}}(\bar{x}_{i-2}))}{\partial x_{i-1}}\right| \beta_{i-1}^{q_i}(\bar{x}_{i-1}) + \left|\frac{\partial \beta_{i-1}^{q_i}}{\partial x_{i-1}}\right| |\xi_{i-1}| \\
&\leq \left|\frac{\partial x_{i-1}^{q_{i-1}}}{\partial x_{i-1}} - 0\right| \beta_{i-1}^{q_i}(\bar{x}_{i-1}) + \left|\frac{\partial \beta_{i-1}^{q_i}(\bar{x}_{i-1})}{\partial x_{i-1}}\right| |\xi_{i-1}| \\
&\leq q_{i-1}\,|x_{i-1}|^{q_{i-1}-1}\,\beta_{i-1}^{q_i}(\bar{x}_{i-1}) + \left|\frac{\partial \beta_{i-1}^{q_i}(\bar{x}_{i-1})}{\partial x_{i-1}}\right| |\xi_{i-1}| \\
&\stackrel{\Delta}{=} B_{i-1}(\bar{x}_{i-1})\,|x_{i-1}|^{q_{i-1}-1} \\
&\quad + \sum_{l=0}^{0} |\xi_{i-1+l}| \left|\frac{\partial \beta_{i-1+l}^{q_{i+l}}(\bar{x}_{i-1+l})}{\partial x_{i-1}}\right| \tilde{\alpha}_{i+l} \text{ with } \tilde{\alpha}_i = 1.
\end{aligned} \tag{3.88}$$

Thus, for $j = 1, 2, \ldots, i-2, i-1$, we can put (3.87) and (3.88) into a general form as follows:

$$\left|\frac{\partial(x_i^{*q_i})}{\partial x_j}\right| \leq B_j(\bar{x}_j)\,|x_j|^{q_j-1} + \sum_{l=0}^{i-j-1} |\xi_{j+l}| \left|\frac{\partial \beta_{j+l}^{q_{j+l+1}}(\bar{x}_{j+l})}{\partial x_j}\right| \tilde{\alpha}_{j+l+1}. \tag{3.89}$$

Furthermore,

$$\begin{aligned}
|x_j|^{q_j-1} &= \left|x_j^{q_j}\right|^{1-\frac{1}{q_j}} = \left|x_j^{q_j} - x_j^{*q_j} + x_j^{*q_j}\right|^{1-\frac{1}{q_j}} \\
&\leq \left[\left|x_j^{q_j} - x_j^{*q_j}\right| + \left|x_j^{*q_j}\right|\right]^{1-\frac{1}{q_j}} \\
&= \left[|\xi_j| + \beta_{j-1}^{q_j}(\bar{x}_{j-1})\,|\xi_{j-1}|\right]^{1-\frac{1}{q_j}}.
\end{aligned} \tag{3.90}$$

By using (1.8), (1.10) and (3.5), it is easy to show that

$$\begin{aligned}
\left|x_{j+1}^{r_{j,k}}\right| &= \left|x_{j+1}^{r_{j,k}} - x_{j+1}^{*r_{j,k}} + x_{j+1}^{*r_{j,k}}\right| \\
&\leq \left|x_{j+1}^{r_{j,k}} - x_{j+1}^{*r_{j,k}}\right| + \left|x_{j+1}^{*r_{j,k}}\right| \\
&= \left|(x_{j+1}^{q_{j+1}})^{\frac{r_{j,k}}{q_{j+1}}} - (x_{j+1}^{*q_{j+1}})^{\frac{r_{j,k}}{q_{j+1}}}\right| + \left|-\beta_j \xi_j^{\frac{r_{j,k}}{q_{j+1}}}\right| \\
&\leq 2^{1-\frac{r_{j,k}}{q_{j+1}}} \left|x_{j+1}^{q_{j+1}} - x_{j+1}^{*q_{j+1}}\right|^{\frac{r_{j,k}}{q_{j+1}}} + \beta_j\,|\xi_j|^{\frac{r_{j,k}}{q_{j+1}}} \\
&= 2^{1-\frac{r_{j,k}}{q_{j+1}}} |\xi_{j+1}|^{\frac{r_{j,k}}{q_{j+1}}} + \beta_j\,|\xi_j|^{\frac{r_{j,k}}{q_{j+1}}},
\end{aligned} \tag{3.91}$$

and

$$
\begin{aligned}
|f_{j,k}| &\le (|x_1| + \cdots + |x_j|)\gamma_{j,k} \\
&\le \left[\sum_{l=1}^{j} |x_l - x_l^*| + \sum_{l=2}^{j} |x_l^*|\right]\gamma_{j,k} \\
&\le \left[\sum_{l=1}^{j} 2^{\frac{q_l-1}{q_l}} |\xi_l|^{\frac{1}{q_l}} + \sum_{l=2}^{j} |\xi_{l-1}|^{\frac{1}{q_l}} \beta_{l-1}(\bar{x}_{l-1})\right]\gamma_{j,k} \\
&\le \left[\sum_{l=1}^{j} |\xi_l|^{\frac{r_{j,k}}{q_{j+1}}} + \sum_{l=2}^{j} |\xi_{l-1}|^{\frac{r_{j,k}}{q_{j+1}}}\right]\bar{\gamma}_{j,k}(\bar{x}_j) \\
&\le \hat{\gamma}_{j,k}(\bar{x}_j)\sum_{l=1}^{j} |\xi_l|^{\frac{r_{j,k}}{q_{j+1}}},
\end{aligned}
\tag{3.92}
$$

where $\bar{\gamma}_{j,k}(\bar{x}_j)$, $\hat{\gamma}_{j,k}(\bar{x}_j)$ are C^1 positive functions.

Using (3.91) and (3.92), the last term in (3.85) can be bounded as

$$
\begin{aligned}
&\left|x_{j+1}^{r_{j,k}} + f_{j,k}\right| \le \left|x_{j+1}^{r_{j,k}}\right| + |f_{j,k}| \\
&\le 2^{1-\frac{r_{j,k}}{q_{j+1}}} |\xi_{j+1}|^{\frac{r_{j,k}}{q_{j+1}}} + \beta_j |\xi_j|^{\frac{r_{j,k}}{q_{j+1}}} + \hat{\gamma}_{j,k}(\bar{x}_j)\sum_{l=1}^{j} |\xi_l|^{\frac{r_{j,k}}{q_{j+1}}} \\
&\le \breve{\gamma}_{j+1,k}(\bar{x}_{j+1})\sum_{l=1}^{j+1} |\xi_l|^{\frac{r_{j,k}}{q_{j+1}}}.
\end{aligned}
\tag{3.93}
$$

From (3.90) and (3.93), it follows that

$$
\begin{aligned}
&B_j[|\xi_j| + \beta_{j-1}^{q_j}|\xi_{j-1}|]^{1-\frac{1}{q_j}} \left|x_{j+1}^{r_{j,k}} + f_{j,k}\right| \\
&\le B_j(|\xi_j| + \beta_{j-1}^{q_j}|\xi_{j-1}|)^{1-\frac{1}{q_j}} \breve{\gamma}_{j+1,k}\sum_{l=1}^{j+1} |\xi_l|^{\frac{r_{j,k}}{q_{j+1}}} \\
&\le \sum_{l=1}^{j+1}\left[\frac{1-\frac{1}{q_j}}{1-\frac{1}{q_j}+\frac{r_{j,k}}{q_{j+1}}}\left(|\xi_j| + \beta_{j-1}^{q_j}|\xi_{j-1}|\right)^{1-\frac{1}{q_j}+\frac{r_{j,k}}{q_{j+1}}}\right. \\
&\quad \left. + \frac{\frac{r_{j,k}}{q_{j+1}}}{1-\frac{1}{q_j}+\frac{r_{j,k}}{q_{j+1}}} |\xi_l|^{1-\frac{1}{q_j}+\frac{r_{j,k}}{q_{j+1}}}\right] B_j\breve{\gamma}_{j+1,k} \\
&\le \sum_{l=1}^{j+1}[\left(|\xi_j| + \beta_{j-1}^{q_j}|\xi_{j-1}|\right)^{1-\frac{1}{q_j}+\frac{r_{j,k}}{q_{j+1}}} + |\xi_l|^{1-\frac{1}{q_j}+\frac{r_{j,k}}{q_{j+1}}}] \\
&\quad \times\bar{\gamma}_{j+1,k}(\bar{x}_{j+1})
\end{aligned}
$$

$$
\begin{aligned}
&\le \sum_{l=1}^{j+1} [2^{\frac{r_{j,k}}{q_{j+1}}-\frac{1}{q_j}} (|\xi_j^{1-\frac{1}{q_j}+\frac{r_{j,k}}{q_{j+1}}}| + |\beta_{j-1}^{q_j}\xi_{j-1}|^{1-\frac{1}{q_j}+\frac{r_{j,k}}{q_{j+1}}}) \\
&\quad + |\xi_l|^{1-\frac{1}{q_j}+\frac{r_{j,k}}{q_{j+1}}}] \times \bar{\gamma}_{j+1,k}(\bar{x}_{j+1}) \\
&\le \sum_{l=1}^{j+1} |\xi_l|^{1-\frac{1}{q_j}+\frac{r_{j,k}}{q_{j+1}}} \tilde{\gamma}_{j+1,l,k}(\bar{x}_{j+1}),
\end{aligned} \tag{3.94}
$$

where $\breve{\gamma}_{j,k}(\bar{x}_j)$, $\bar{\gamma}_{j,l,k}(\bar{x}_j)$ and $\tilde{\gamma}_{j,l,k}(\bar{x}_j)$ are C^1 positive functions. Note that (1.9) and the relation $1 \ge 1 - \frac{1}{q_j} + \frac{r_{j,k}}{q_{j+1}} \in Q_{odd}$ are used in the last inequality.

By using (1.10), from (3.89) and (3.93), we obtain

$$
\begin{aligned}
&\sum_{l=0}^{i-j-1} |\xi_{j+l}| \left| \frac{\partial \beta_{j+l}^{q_{j+l+1}}(\bar{x}_{j+l})}{\partial x_j} \right| \tilde{\alpha}_{j+l+1} \left| x_{j+1}^{r_{j,k}} + f_{j,k} \right| \\
&\le \sum_{l=0}^{i-j-1} |\xi_{j+l}|^{1-\frac{1}{q_j}+\frac{r_{j,k}}{q_{j+1}}} \lambda_{2,j}(\bar{x}_{j+1}),
\end{aligned} \tag{3.95}
$$

where $\lambda_{2,j}(\bar{x}_{j+1})$ are positive C^1 functions.
With the help of (3.94) and (3.95), we have

$$
\begin{aligned}
&\left| \frac{\partial (x_i^{*q_i})}{\partial x_j} \right| \left| x_{j+1}^{r_{j,k}} + f_{j,k}(\bar{x}_j) \right| \\
&\le \sum_{l=1}^{j+1} |\xi_l|^{1-\frac{1}{q_j}+\frac{r_{j,k}}{q_{j+1}}} \tilde{\gamma}_{j,l,k}(\bar{x}_j) + \sum_{l=0}^{i-j-1} |\xi_{j+l}|^{1-\frac{1}{q_j}+\frac{r_{j,k}}{q_{j+1}}} \lambda_{2,j}(\bar{x}_{j+1}) \\
&\le \sum_{l=1}^{j+1} |\xi_l|^{1-\frac{1}{q_j}+\frac{r_{j,k}}{q_{j+1}}} \lambda_{3,j}(\bar{x}_{j+1}),
\end{aligned} \tag{3.96}
$$

where $\lambda_{3,j}(\bar{x}_{j+1})$ are positive C^1 functions.

Using (1.9) and Assumption 3.1, and substituting (3.70) and (3.96) into (3.85), it follows that

$$
\begin{aligned}
&\sum_{j=1}^{i-1} \frac{\partial W_i(\bar{x}_i)}{\partial x_j} \dot{x}_j \\
&\le \sum_{j=1}^{i-1} (2 - 1/q_i) 2^{1-\frac{1}{q_i}} |\xi_i| \sum_{l=1}^{j+1} |\xi_l|^{1-\frac{1}{q_j}+\frac{r_{j,k}}{q_{j+1}}} \lambda_{3,j}(\bar{x}_{j+1}) \\
&= \sum_{j=1}^{i-1} \sum_{l=1}^{j+1} (2 - 1/q_i) 2^{1-\frac{1}{q_i}} |\xi_i| \, |\xi_l|^{1-\frac{1}{q_j}+\frac{r_{j,k}}{q_{j+1}}} \lambda_{3,j}(\bar{x}_{j+1})
\end{aligned}
$$

$$\leq \frac{1}{3}\sum_{j=1}^{i-1} \xi_j^{1+\tau_k} + c_{i,3,k}(\bar{x}_i)\xi_i^{1+\tau_k}, \tag{3.97}$$

where $c_{i,3,k}(\bar{x}_i)$ is a positive C^1 function. Thus, the existence of the functions $c_{i,3,k}(\bar{x}_i)$ is proved. The proof of proposition is completed.

References

1. Long, L., Zhao, J.: H_∞ control of switched nonlinear systems in p-normal form using multiple lyapunov functions. IEEE Trans. Autom. Control **57**(5), 1285–1291 (2012)
2. Ma, R., Liu, Y., Zhao, S., Wang, M., Zong, G.: Global stabilization design for switched power integrator triangular systems with different powers. Nonlinear Anal. Hybrid Syst. **15**(2), 74–85 (2015)
3. Long, L., Zhao, J.: Global stabilisation of switched nonlinear systems in p-normal form with mixed odd and even powers. Int. J. Control **84**(10), 1612–1626 (2011)
4. Hespanha, J.P., Morse, A.S.: Switching between stabilizing controllers. Automatica **38**(11), 1905–1917 (2002)
5. Branicky, M.S.: Multiple lyapunov functions and other analysis tools for switched and hybrid systems. IEEE Trans. Autom. Control **43**(4), 475–482 (1998)
6. Liberzon, D., Morse, A.S.: Basic problems in stability and design of switched systems. IEEE Control Syst. Mag. **19**(5), 59–70 (1999)
7. Lin, H., Antsaklis, P.J.: Stability and stabilizability of switched linear systems: a survey of recent results. IEEE Trans. Autom. Control **54**(2), 308–322 (2009)
8. Vu, L., Liberzon, D.: Invertibility of switched linear systems. Automatica **44**(4), 949–958 (2008)
9. Zhao, J., Hill, D.J.: On stability, L_2-gain and H_∞ control for switched systems. Automatica **44**(5), 1220–1232 (2008)
10. Liu, J., Liu, X., Xie, W.-C.: Class-*kl* estimates and input-to-state stability analysis of impulsive switched systems. Syst. Control Lett. **61**(6), 738–746 (2012)
11. Bras, I., Carapito, A.C., Rocha, P.: Stability of switched systems with partial state reset. IEEE Trans. Autom. Control **58**(4), 1008–1012 (2013)
12. Back, J., Cheong, S.G., Shim, H., Seo, J.H.: Nonsmooth feedback stabilizer for strict-feedback nonlinear systems that may not be linearizable at the origin. Syst. Control Lett. **56**(11-12), 742–752 (2007)
13. Ding, S., Li, S.: Global finite-time stabilization of nonlinear integrator systems subject to input saturation. ACTA Automatica Sinica **37**(10), 1222–1231 (2011)
14. Han, T.-T., Ge, S.S., Lee, H.T.: Adaptive neural control for a class of switched nonlinear systems. Syst. Control Lett. **58**(2), 109–118 (2009)
15. Ma, R., Zhao, J.: Backstepping design for global stabilization of switched nonlinear systems in lower triangular form under arbitrary switchings. Automatica **46**(11), 1819–1823 (2010)
16. Wu, J.-L.: Stabilizing controllers design for switched nonlinear systems in strict-feedback form. Automatica **45**(4), 1092–1096 (2009)
17. Ames, A.D., Zheng, H., Gregg, R.D., Sastry, S., IEEE.: Is there life after zeno? taking executions past the breaking (zeno) point. In: 2006 American Control Conference, pp. 2652–2657 (2006)
18. Yu, L., Barbot, J.P., Boutat, D., Benmerzouk, D., IEEE.: Observability normal forms for a class of switched systems with zeno phenomena. In: 2009 American Control Conference, pp. 1766–1771 (2009)
19. Xu, X., Antsaklis, P.J.: A dynamic programming approach for optimal control of switched systems. In: Proceedings of the 39th IEEE Conference on Decision and Control, pp. 1822–1827 (2000)

20. Qian, C., Lin, W.: Non-lipschitz continuous stabilizers for nonlinear systems with uncontrollable unstable linearization. Syst. Control Lett. **42**(3), 185–200 (2001)
21. Khalil, H.K.: Nonlinear Systems. Prentice Hall, New Jersey (2000)
22. Huang, X., Lin, W., Yang, B.: Global finite-time stabilization of a class of uncertain nonlinear systems. Automatica **41**(5), 881–888 (2005)

Chapter 4
Global Finite-Time Stabilization of Switched Nonlinear Systems in Non-triangular Form

This chapter investigates the global finite-time stabilization for a class of switched nonlinear systems in non-triangular form, whose subsystems have chained integrators with the powers of positive odd rational numbers (i.e., numerators and denominators of the powers are all positive odd integers). All subsystems are not assumed to be stabilizable. Based on the technique of adding a power integrator and the multiple Lyapunov functions method, both the global finite-time stabilizers of individual subsystems and a switching law are systematically constructed to guarantee global finite-time stabilization of the closed-loop switched nonlinear system. A numerical example is provided to illustrate the effectiveness of the proposed method.

4.1 Introduction

In previous two chapters, the finite-time stabilization of switched systems has been investigated. One feature of the studied switched systems is that all subsystems are with special structure in lower triangular form. It is known that constructive technique for finite-time stabilization is one of the active research topics of the modern control theory. On the other hand, by designing a suitable switching sequence, or switching law, one can stabilize a switched system when all subsystems are unstable. Therefore, with considerable effort, is it possible to extend the existing constructive procedures and switching signal design method to wider classes of switched nonlinear systems in, for example, non-triangular form.

Several asymptotical stabilization results have been obtained by utilizing structural characteristics of switched systems. The stabilization for high-order switched systems has been studied in [1] by adding a power integrator technique. For switched systems, if the system structure is in non-triangular form, but in special structure, [2] investigated the stabilization of switched systems by designing an appropriate switching law and individual subsystem controllers. However, the powers of inte-

J. Fu and R. Ma, *Stabilization and H_∞ Control of Switched Dynamic Systems*, Studies in Systems, Decision and Control 310,
https://doi.org/10.1007/978-3-030-54197-2_4

grators of switched systems studied in [1, 2] can be only positive integers, and thus these methods can not deal with the switched systems where powers of the chained integrators are not integers. *Is it possible to global finite-time stabilize switched systems whose powers of integrators are not necessarily integers?* On the other hand, it is very interesting to investigate the finite-time stabilization for switched systems, the structure of whose subsystems is more general, which is no longer the one in [1, 3–6] but contain them as a special case. If possible, under what conditions can we design such a controller and how? To our best knowledge, in the literature there have not been results which provide answers to these questions.

This chapter will study the global finite-time stabilization for a class of switched systems with the following distinguished characteristics: (i) The powers of the chained integrators of all subsystems are positive odd rational numbers (i.e., numerators and denominators of the powers are all positive odd integers). Since the powers can be less than 1, switched system does not have to have first approximation at the origin, which makes the standard backstepping not applicable. (ii) The switched systems are in non-triangular form but contain strict-feedback form, power integrator triangular form, etc. as a special form. (iii) All subsystems are not assumed to be asymptotically stabilizable. Therefore, backstepping technique seems to be not utilized to such a studied switched system since backstepping is restricted in triangular form. However, to our best knowledge, in the literature there have not been results which provide answers to the global finite-time stabilization problem for such a switched system, which also partially motivates our present work. To address these challenge problems, we will design controllers for individual subsystems and construct a switching law to achieve global finite-time stabilization of the closed-loop switched systems. Compared to the relevant existing results in the literature, the main contributions of this paper are as follows:

- The global finite-time stabilization of switched nonlinear system, where the powers of the chained integrators of $\dot{x}_i$-equations can be positive odd rational numbers and whose subsystems are in non-triangular, is obtained for the first time.
- Under the assumption that all subsystems are not asymptotically stabilizable, sufficient conditions for the global finite-time stabilization are derived.
- Both a recursive design algorithm for constructing individual stabilizer of switched system and a switching law are presented.

4.2 Preliminaries and System Description

We consider the switched nonlinear systems described by the following equations:

$$\begin{aligned}
\dot{x}_i &= x_{i+1}^{r_{i,\sigma(t)}} + f_{i,\sigma(t)}(\bar{x}_i) + g_{i,\sigma(t)}(x), \\
i &= 1, \ldots, n-1, \\
\dot{x}_n &= u^{r_{n,\sigma(t)}} + f_{n,\sigma(t)}(x) + g_{n,\sigma(t)}(x),
\end{aligned} \tag{4.1}$$

where $x = (x_1, \ldots, x_n) \in R^n$ and $u \in R$ are system state and control input, respectively, $\bar{x}_i = (x_1, \ldots, x_i) \in R^i, i = 1, 2, \ldots, n$; the function $\sigma(t)$ is the switching signal which is assumed to be a piecewise continuous (from the right) function depending on time or state or both. m is the number of models (called subsystems) of the switched system. For any $k \in M$, and $i = 1, 2, \ldots, n$, $f_{i,k}(x_1, \ldots, x_i) : R^+ \to R$ and $g_{i,k}(x) : R^n \to R$, are C^1 and vanish at the origin, and $r_{i,k}$s are rational numbers whose numerators and denominators are all positive odd integers (we will call such $r_{i,k}$ a positive odd rational number), and satisfy the following assumption:

Assumption 4.1 There are constants $q_{i,k} \in Q_{odd}$ and $q_{i,k} \geq 1, i = 1, 2, \ldots, n+1$, such that, $i = 1, 2, \ldots, n, k \in M$,

$$q_{1,k} = 1, \quad 1 + \frac{r_{i,k}}{q_{i+1,k}} = \tau_k + \frac{1}{q_{i,k}}, \tag{4.2}$$

$$\frac{r_{i,k}}{q_{i+1,k}} \leq \min\left\{\frac{1}{q_{1,k}}, \frac{1}{q_{2,k}}, \ldots, \frac{1}{q_{i,k}}\right\}, \tag{4.3}$$

where $\tau_k \in (0, 1)$ and $\tau_k \in Q_{odd}$.

The terms $f_{n,\sigma(t)}(x) + g_{n,\sigma(t)}(x)$ in $\dot{x}_n$ can be combined into one. However, to simplify the following description, we still have this term $g_{n,\sigma(t)}(x)$, but this one is always zero.

Our objective is to solve the global finite-time stabilization problem for system (4.1), that is, under what kind of conditions, are there individual controller for subsystems and an appropriate switching law which render $x = 0$ of (4.1) finite-time stable, and how to design systematically them if they are. If at least one subsystem is asymptotically stabilizable based on the existing methods, this problem is trivial. Therefore, none of the individual subsystems is assumed to be asymptotically stabilizable.

Remark 4.1 The system (4.1) represents a switched nonlinear system in non-triangular form, since $g_{i,k}(x), 1 \leq i \leq n, k \in M$, are functions of all the state variables, each subsystem of which is much more general than the structure of the non-switched nonlinear systems in [7]. Specifically, when the subsystem of (4.1) involves no nonlinearities $g_{i,k}(x)$, that is, $g_{i,k}(x) \equiv 0$, it becomes the one studied in [7]. One the other hand, from the perspective of the powers, the powers of the chained integrators of switched nonlinear systems can be positive odd rational numbers, which contains the positive odd integer as a special case.

4.3 Main Result

In this section, we first construct a set of Lyapunov functions for all subsystems of the following switched systems:

$$\begin{aligned}
\dot{x}_i &= x_{i+1}^{r_{i,\sigma(t)}} + f_{i,\sigma(t)}(\bar{x}_i), \\
i &= 1, \ldots, n-1, \\
\dot{x}_n &= u^{r_{n,\sigma(t)}} + f_{n,\sigma(t)}(x),
\end{aligned} \tag{4.4}$$

and then present a global finite-time stabilization result for the system (4.1) by constructing controllers for subsystems and a switching law under some appropriate conditions.

Let $f_{i,k} : R^i \to R$ be C^1 with $f_{i,k}(0) = 0$. Then, there exists a C^1 nonnegative function $\gamma_{i,k}(\bar{x}_i)$ such that

$$\left| f_{i,k}(\bar{x}_i) \right| \le (|x_1| + \cdots + |x_i|)\gamma_{i,k}(\bar{x}_i). \tag{4.5}$$

We now design individual controller for subsystem k of (4.4).

Step 1. For the first step of the induction, choose $V_{1,k}(x_1) = \frac{1}{2}x_1^2$ and a C^1 function $L_{1,k}(x_1)$. For subsystem k in (4.4), by using (4.5), a simple computation gives

$$\begin{aligned}
\dot{V}_{1,k} &= x_1 x_1^{r_{1,k}} + x_1 f_{1,k}(x_1) \\
&= x_1(x_1^{r_{1,k}} - x_2^{*r_{1,k}}) + x_1 x_2^{*r_{1,k}} + x_1 f_{1,k}(x_1) \\
&\le x_1(x_1^{r_{1,k}} - x_2^{*r_{1,k}}) + x_1 x_2^{*r_{1,k}} + x_1^2 \gamma_{1,k}(x_1),
\end{aligned} \tag{4.6}$$

where x_2^* is the virtual controller to be designed.

With the help of (1.10) and $0 < \frac{r_{1,k}}{q_{2,k}} < 1$, one gets

$$|x_1| \le |x_1|^{\frac{r_{1,k}}{q_{2,k}}} + |x_1| = |x_1|^{\frac{r_{1,k}}{q_{2,k}}}(1 + |x_1|^{1-\frac{r_{1,k}}{q_{2,k}}}). \tag{4.7}$$

By taking a C^1 function $\tilde{\gamma}_{1,k}(x_1)$ such that $(1 + |x_1|^{1-\frac{r_{1,k}}{q_{2,k}}})\gamma_{1,k}(x_1) \le \tilde{\gamma}_{1,k}(x_1)$, then we obtain

$$\dot{V}_1 \le x_1(x_1^{r_{1,k}} - x_{2,k}^{*r_{1,k}}) + x_1 x_{2,k}^{*r_{1,k}} + |x_1|^{1+\frac{r_{1,k}}{q_{2,k}}} \tilde{\gamma}_{1,k}. \tag{4.8}$$

For the first virtual controller $x_{2,k}^*$, we take $x_{2,k}^*$ as:

$$x_{2,k}^* = -\beta_{1,k}(x_1) x_1^{\frac{1}{q_{2,k}}} \tag{4.9}$$

with a C^1 function $\beta_{1,k}(x_1) = (n + L_{1,k}(x_1)(1 + x_1^2) + \tilde{\gamma}_{1,k}(x_1))^{1/r_{1,k}}$. Substituting (4.9) into (4.8) yields

$$\dot{V}_{1,k} \le -n x_1^{1+\tau_k} - L_{1,k}(x_1) x_1^2 + x_1(x_2^{r_{1,k}} - x_{2,k}^{*r_{1,k}}). \tag{4.10}$$

Inductive step. For step i, we use induction. Suppose at step $i-1$, there is a C^1 Lyapunov function $V_{i-1,k}(\xi_{i-1,k})$, which is positive definite and proper, satisfying

$$V_{i-1,k}(\bar{\xi}_{i-1,k}) \leq 2\left(\xi_{1,k}^2 + \cdots + \xi_{i-1,k}^2\right), \tag{4.11}$$

and a set of C^0 virtual controllers $x_{1,k}^*, \ldots, x_{i,k}^*$ defined by

$$\begin{aligned} x_{1,k}^* &= 0, & \xi_{1,k} &= x_1 - x_{1,k}^*, \\ x_{2,k}^* &= -\beta_{1,k}(\bar{\xi}_{1,k})\xi_{1,k}^{\frac{1}{q_{2,k}}}, & \xi_{2,k} &= x_2^{q_{2,k}} - x_{2,k}^{*q_{2,k}}, \\ &\vdots & &\vdots \\ x_{i,k}^* &= -\beta_{i-1,k}(\bar{\xi}_{i-1,k})\xi_{i-1,k}^{\frac{1}{q_{i,k}}}, & \xi_{i,k} &= x_i^{q_{i,k}} - x_{i,k}^{*q_{i,k}}, \end{aligned} \tag{4.12}$$

with a set of functions $\beta_{j,k}(\bar{\xi}_{j,k}) > 0$, $L_{j,k}(\bar{\xi}_{j,k}) > 0$, $1 \leq j \leq i-1$, such that, when $\nu_{i-1,k} = 2 - \frac{1}{q_{i-1,k}}$

$$\begin{aligned} \dot{V}_{i-1,k}(\bar{\xi}_{i-1,k}) \leq & -(n-i+2)\sum_{j=1}^{i-1}\xi_{j,k}^{1+\tau_k} - \sum_{j=1}^{i-1}L_{j,k}(\bar{\xi}_{j,k})\xi_{j,k}^2 \\ & + \xi_{i-1,k}^{\nu_{i-1,k}}(x_i^{r_{i-1,k}} - x_{i,k}^{*r_{i-1,k}}). \end{aligned} \tag{4.13}$$

We aim to construct $V_{i,k}(\bar{\xi}_{i,k})$ and $x_{i+1,k}^*$, such that the analog of (4.11) and (4.13) hold with i replacing $i-1$. To prove this claim, we consider a Lyapunov function defined by

$$V_{i,k}(\bar{\xi}_{i,k}) = V_{i-1,k}(\bar{\xi}_{i-1,k}) + W_{i,k}(\bar{\xi}_{i,k}), \tag{4.14}$$

where

$$W_{i,k} = \int_{x_{i,k}^*}^{x_i} (s^{q_{i,k}} - x_{i,k}^{*q_{i,k}})^{\nu_{i,k}} ds, \; \nu_{i,k} = 2 - \frac{1}{q_{i,k}}. \tag{4.15}$$

By using a similar method as in [7], it is easy to show that $V_{i,k}(\bar{\xi}_{i,k})$ is C^1, positive definite and proper function, and $W_{i,k}(\bar{\xi}_{i,k})$ is C^1 with the following two important properties: for $j = 1, 2, \ldots, i-1$,

$$\frac{\partial W_{i,k}(\bar{\xi}_{i,k})}{\partial x_j} = -\nu_{i,k}\int_{x_i^*}^{x_i}(s^{q_{i,k}} - x_i^{*q_{i,k}})^{\nu_{i,k}-1} ds \frac{\partial x_i^{*q_{i,k}}}{\partial x_j}, \tag{4.16}$$

$$\frac{\partial W_{i,k}(\bar{\xi}_{i,k})}{\partial x_i} = (x_i^{q_{i,k}} - x_{i,k}^{*q_{i,k}})^{2-\frac{1}{q_{i,k}}} = \xi_{i,k}^{2-\frac{1}{q_{i,k}}}. \tag{4.17}$$

From (4.4), (4.13)–(4.17), a straightforward calculation gives

$$\dot{V}_{i,k} \leq -(n-i+2)\sum_{j=1}^{i-1}\xi_{j,k}^{1+\tau_k} - \sum_{j=1}^{i-1}L_{j,k}\xi_{j,k}^2$$

$$+\xi_{i-1,k}^{2-\frac{1}{q_{i-1,k}}}(x_i^{r_{i-1,k}} - x_{i,k}^{*r_{i-1,k}}) + \sum_{j=1}^{i-1}\frac{\partial W_{i,k}}{\partial x_j}\dot{x}_j$$
$$+\xi_{i,k}^{2-\frac{1}{q_{i,k}}}(x_{i+1}^{r_{i,k}} + f_{i,k}(\bar{x}_i)). \tag{4.18}$$

We estimate the last three terms on the right-hand side of (4.18).
One can obtain that

$$\begin{aligned} | x_i^{r_{i-1,k}} - x_{i,k}^{*r_{i-1,k}} | &\le | x_i^{q_{i,k}} - x_{i,k}^{*q_{i,k}} |^{\frac{r_{i-1,k}}{q_{i,k}}} 2^{1-\frac{r_{i-1,k}}{q_{i,k}}} \\ &= 2^{1-\frac{r_{i-1,k}}{q_{i,k}}} | \xi_{i,k}^{\frac{r_{i-1,k}}{q_{i,k}}} | . \end{aligned} \tag{4.19}$$

Thus, by (1.8), (1.9), and (4.19), there exist constants $c_{i,1,k}$, $k = 1, 2, \ldots, m$, such that

$$\begin{aligned} &\left|\xi_{i-1,k}^{2-\frac{1}{q_{i-1,k}}}(x_i^{r_{i-1,k}} - x_{i,k}^{*r_{i-1,k}})\right| \\ &\le 2^{1-\frac{r_{i-1,k}}{q_{i,k}}} \left|\xi_{i-1,k}\right|^{2-\frac{1}{q_{i-1,k}}} \left|\xi_{i,k}\right|^{\frac{r_{i-1,k}}{q_{i,k}}} \\ &= \frac{1}{3}(\xi_{1,k}^{1+\tau_k} + \cdots + \xi_{i-1,k}^{1+\tau_k}) + c_{i,1,k}\xi_{i,k}^{1+\tau_k}. \end{aligned} \tag{4.20}$$

We introduce two propositions for the sake of fulfilling our goal, which are useful to estimate the last two terms of (4.18).

Proposition 4.2 *There exist C^1 functions $c_{i,2,k}(\bar{\xi}_i) \ge 0$, $k = 1, 2, \ldots, m$, such that*

$$\sum_{j=1}^{i-1}\frac{\partial W_{i,k}}{\partial x_j}\dot{x}_j \le \frac{1}{3}\sum_{j=1}^{i-1}\xi_{j,k}^{1+\tau_k} + c_{i,2,k}(\bar{\xi}_{i,k})\xi_{i,k}^{1+\tau_k}. \tag{4.21}$$

Proposition 4.3 *There exist C^1 functions $c_{i,3,k}(\bar{\xi}_{i,k}) \ge 0$, $k = 1, 2, \ldots, m$, such that*

$$\left|\xi_{i,k}^{2-\frac{1}{q_{i,k}}} f_{i,k}(\bar{x}_i)\right| \le \frac{1}{3}\sum_{j=1}^{i-1}\xi_{j,k}^{1+\tau_k} + c_{i,3,k}(\bar{\xi}_{i,k})\xi_{i,k}^{1+\tau_k}. \tag{4.22}$$

The proof of Proposition 4.2 and 4.3 can be easily proved by using the same method as in [7], and are omitted here.

Now, substituting (4.20), (4.21) and (4.22) into (4.18) yields

$$\begin{aligned} \dot{V}_{i,k} \le &-(n-i+1)\sum_{j=1}^{i-1}\xi_{j,k}^{1+\tau_k} - \sum_{j=1}^{i-1}L_{j,k}(\bar{\xi}_{j,k})\xi_{j,k}^2 \\ &+(c_{i,1,k} + c_{i,2,k}(\bar{\xi}_{i,k}) + c_{i,3,k}(\bar{\xi}_{i,k}))\xi_{i,k}^{1+\tau_k} \end{aligned}$$

$$+\xi_{i,k}^{2-\frac{1}{q_{i,k}}} x_{i+1}^{*r_{i,k}} + \xi_{i,k}^{2-\frac{1}{q_{i,k}}} (x_{i+1}^{r_{i,k}} - x_{i+1,k}^{*r_{i,k}}). \tag{4.23}$$

From Assumption 4.1, we have $2 - \frac{1}{q_{i,k}} + \frac{r_{i,k}}{q_{i+1,k}} = 1 + \tau_k$. Therefore, by taking the virtual control $x_{i+1,k}^*$ as

$$x_{i+1,k}^* = -\beta_{i,k}(\bar{\xi}_{i,k})\xi_{i,k}^{\frac{1}{q_{i+1,k}}} \tag{4.24}$$

with C^1 function $\beta_{i,k}(\bar{\xi}_{i,k})=(c_{i,1,k} + c_{i,2,k}(\bar{\xi}_{i,k}) + c_{i,3,k}(\bar{\xi}_{i,k}) + L_{j,k}(\bar{\xi}_{j,k})(1 + \xi_{i,k}^2) + n - i + 1)^{\frac{1}{r_{i,k}}}$, one gets

$$\dot{V}_{i,k} \le -(n - i + 1)\sum_{j=1}^{i} \xi_{j,k}^{1+\tau_k} - \sum_{j=1}^{i} L_{j,k}(\bar{\xi}_{j,k})\xi_{j,k}^2 + \xi_{i,k}^{2-\frac{1}{q_{i,k}}} (x_{i+1}^{r_{i,k}} - x_{i+1,k}^{*r_{i,k}}). \tag{4.25}$$

Inequality (4.25) is the desired ith step analog of inequality (4.13), which proves the induction. Note again that x_{i+1}^* and $x_{i+1}^{*r_{i,k}}$ may not be C^1, but $x_{i+1}^{*q_{i+1}}$ is.

Using the inductive argument above, one concludes that at step n, there is a C^1 state feedback controller of the form

$$u_k = x_{n+1,k}^* = -\beta_{n,k}(\bar{\xi}_{n,k})\xi_{n,k}^{\frac{1}{q_{n+1,k}}}, \tag{4.26}$$

with the C^1 proper and positive definite Lyapunov function $V_{n,k}(x)$ constructed via the inductive procedure, we arrive at

$$\dot{V}_{n,k}(\bar{\xi}_{n,k}) \le -\sum_{j=1}^{n} \xi_{j,k}^{1+\tau_k} - \sum_{j=1}^{n} L_{j,k}(\bar{\xi}_{j,k})\xi_{j,k}^2. \tag{4.27}$$

Recall that $V_{n,k}(\bar{\xi}_{n,k}) = \sum_{j=1}^{n} W_{j,k}(\bar{\xi}_{j,k})$ in (4.14). When constants $\lambda_k : 0 < \lambda_k < 1$, it is easy to compute that

$$V_{n,k}^{\lambda_k}(\bar{\xi}_{n,k}) = [\sum_{j=1}^{n} W_{j,k}(\bar{\xi}_{j,k})]^{\lambda_k} \le \sum_{j=1}^{n} W_{j,k}^{\lambda_k}(\bar{\xi}_{j,k}). \tag{4.28}$$

Moreover, we can obtain that

$$
\begin{aligned}
W_{j,k}(\bar{\xi}_{j,k}) &= \int_{x_{j,k}^*}^{x_j} \left(s^{q_{j,k}} - x_{j,k}^{*q_{j,k}}\right)^{2-\frac{1}{q_{j,k}}} ds \\
&\le \left|x_j - x_{j,k}^*\right| \left|x_j^{q_{j,k}} - x_{j,k}^{*q_{j,k}}\right|^{2-\frac{1}{q_{j,k}}} \\
&= \left[\left|x_j - x_{j,k}^*\right|^{q_{j,k}}\right]^{\frac{1}{q_{j,k}}} \left|x_j^{q_{j,k}} - x_{j,k}^{*q_{j,k}}\right|^{2-\frac{1}{q_{j,k}}} \\
&\le [2^{q_{j,k}-1} \mid x_j^{q_{j,k}} - x_{j,k}^{*q_{j,k}} \mid]^{\frac{1}{q_{j,k}}} \mid x_j^{q_{j,k}} - x_{j,k}^{*q_{j,k}} \mid^{2-\frac{1}{q_{j,k}}} \\
&\le 2^{\frac{q_{j,k}-1}{q_{j,k}}} \mid x_j^{q_{j,k}} - x_{j,k}^{*q_{j,k}} \mid^{\frac{1}{q_{j,k}}+2-\frac{1}{q_{j,k}}} \le 2 \mid \xi_{j,k} \mid^2 .
\end{aligned} \tag{4.29}
$$

That is, $W_{j,k}(\bar{\xi}_{j,k}) \le 2\left|\xi_{j,k}\right|^2$. Thus, one has

$$
\begin{aligned}
V_{n,k}(x) &\le \frac{1}{2}x_1^2 + \left|x_1 - x_{1,k}^*\right| \left|\xi_{2,k}\right|^{2-\frac{1}{q_{n,k}}} + \cdots + \left|x_n - x_{n,k}^*\right| \left|\xi_{n,k}\right|^{2-\frac{1}{q_{n,k}}} \\
&\le 2\sum_{j=1}^{n} \xi_{j,k}^2 .
\end{aligned} \tag{4.30}
$$

Define $c_k = 2^{-\frac{1+\tau_k}{2}}$. By using (4.27), it is easy to see that

$$
\begin{aligned}
\dot{V}_{n,k}(x) &\le -\sum_{j=1}^{n} (\xi_{j,k}^2)^{\frac{1+\tau_k}{2}} - \sum_{j=1}^{n} L_{j,k}(\bar{\xi}_{j,k})\xi_{j,k}^2 \\
&\le -[\sum_{j=1}^{n} \xi_{j,k}^2]^{\frac{1+\tau_k}{2}} - \sum_{j=1}^{n} L_{j,k}(\bar{\xi}_{j,k})\xi_{j,k}^2 \\
&\le -(\frac{1}{2}V_{n,k})^{\frac{1+\tau_k}{2}} - \sum_{j=1}^{n} L_{j,k}(\bar{\xi}_{j,k})\xi_{j,k}^2 \\
&= -c_k V_{n,k}^{\frac{1+\tau_k}{2}} - \sum_{j=1}^{n} L_{j,k}(\bar{\xi}_{j,k})\xi_{j,k}^2 .
\end{aligned} \tag{4.31}
$$

That is,

$$
\dot{V}_{n,k}(x) + c_k V_{n,k}^{\frac{1+\tau_k}{2}}(x) \le -\sum_{j=1}^{n} L_{j,k}(\bar{\xi}_{j,k})\xi_{j,k}^2 . \tag{4.32}
$$

It is easy to verify that $0 < c_k < 1$. Then, $V_{n,k}(x)$ is the Lyapunov function for the closed-loop subsystem k of switched system (4.3). Thus, from Lemma 1.6, the closed-loop subsystem k is globally finite-time stable.

Remark 4.4 Compared with design method in [7], there is an additional term $L_{i,k}(\bar{\xi}_{i,k})(1 + \xi_{i,k}^2)$ in the above designed virtual controller $x_{i,k}^*$, which leads to an additional term $-\sum_{j=1}^{n} L_{j,k}(\bar{\xi}_{j,k})\xi_{j,k}^2$ in $\dot{V}_{n,k}(x)$. This term may give us considerable freedom in the finite-time stabiliser design. If the function $L_{i,k}(\bar{\xi}_{i,k})$ is reduced to a constant and the powers are 1, it is clear that the obtained design method recovers the global finite-time stabilization result in [8].

Next, we construct controllers for subsystems and a switching law to achieve global finite-time stabilization for the switched system (4.1).

Theorem 4.5 *Consider the switched nonlinear system (4.1) and suppose that there exist C^1 functions $\beta_{j,k}(\bar{x}_j) > 0$, $1 \le j \le n$, $k \in M$, satisfying (4.12), and continuous functions $w_{j,k}(x) > 0$, $j, m \in M$, such that, $\forall x \neq 0, k \in M$,*

$$\begin{aligned}&\sum_{j=1}^{n} L_{j,k}(\bar{\xi}_{j,k})\xi_{j,k}^2 - \sum_{j=1}^{n}\sum_{l=1}^{j}\left(\frac{\partial W_{j,k}}{\partial x_l} g_{l,k}\right) - \sum_{j=1}^{n-1}\frac{\partial W_{n,k}}{\partial x_j} g_{j,k}\\ &-\sum_{j=1}^{m} w_{j,k}(\bar{x}_j)\left(V_{n,j} - V_{n,k}\right) > 0,\end{aligned} \tag{4.33}$$

Then, there exist state feedback controllers (4.26) for subsystems and a switching law

$$\sigma(t) = \max\left\{k|k = \arg\min_{k\in M}\left\{V_{n,k}(x)\right\}\right\} \tag{4.34}$$

such that the closed-loop switched system (4.1) is globally finite-time stable.

Proof The main ingredients of the proof are to construct individual controller for subsystems and a switching law which satisfy the conditions of the MLFs method.

Substituting (4.26) into system (4.1) results in

$$\begin{aligned}&\dot{V}_{n,k}\\ &\le -c_k V_{n,k}^{\frac{1+\tau_k}{2}} - \sum_{j=1}^{n} L_{j,k}(\bar{\xi}_{j,k})\xi_{j,k}^2 + \sum_{j=1}^{n}\sum_{l=1}^{j}\left(\frac{\partial W_{j,k}}{\partial x_l} g_{l,k}\right) + \sum_{j=1}^{n-1}\frac{\partial W_{n,k}}{\partial x_j} g_{j,k}\\ &\le -c_k V_{n,k}^{\frac{1+\tau_k}{2}}\\ &< 0, \forall x \neq 0,\end{aligned} \tag{4.35}$$

where $V_{n,k}$ is used as Lyapunov function for subsystem k.

Consider the time interval $t \in [t_h, t_{h+1}), h = 0, 1, 2, \ldots$. Integrating the time derivative of (4.35), the following inequality is obtained

$$V_{n,k}^{1-s_k}(x(t)) \le V_{n,k}^{1-s_k}(x(t_h)) - c_k(1 - s_k)(t - t_h), \tag{4.36}$$

for all $t \in [t_h, t_{h+1}), h = 0, 1, 2, \ldots,$ where $s_k \triangleq \frac{1+\tau_k}{2}$. Note that (4.36) only holds for $V_{n,k}(x) \geq 0$. Using this inequality, we will prove finite-time convergence to the origin from any initial state $x(0)$. Applying this inequality recursively and taking into account that when $V_{n,k}(x) = 0$, there exists t^* such that $V_{n,k}(x(t)) = 0$ for all $t > t^*$ which implies that $x(t) = 0$ for all $t > t^*$, and the system has converged to the origin and remains there. In this case, finite-time convergence to the origin has been achieved.

In the following, we will show the stability of the origin in the sense of Lyapunov. By (4.35), when $\sigma(t) = k$, one has

$$\dot{V}_{n,k}(x) < 0, \forall x \neq 0. \tag{4.37}$$

Therefore, according to the MLFs method, the closed-loop system (4.1) and (4.26) under the switching law (4.34) is asymptotically stable. In above, global finite-time stabilization of the origin is guaranteed because $V_{n,k}(x)$ is radially unbounded. This completes the proof of Theorem 4.5.

4.4 An Illustrative Example

In this section, we present a numerical example to demonstrate the effectiveness of the proposed design method.

Consider the switched system (4.3) with two subsystems in the case when $n = 2$, i.e.

$$\begin{aligned}\dot{x}_1 &= x_2^{r_{1,\sigma(t)}} + g_{1,\sigma(t)}(x),\\ \dot{x}_2 &= u_{\sigma(t)}^{r_{2,\sigma(t)}},\end{aligned} \tag{4.38}$$

where $\sigma(t) : [0, \infty) \to \{1, 2\}$, and $r_{1,1} = r_{2,1} = \frac{1}{3}, r_{1,2} = r_{2,2} = \frac{1}{5}$.

We will design a state feedback control law for each subsystem and a switching law to globally finite-time stabilize the switched system (4.38).

One can verify that Assumption 4.1 holds with $q_1 = q_2 = q_3 = 1, \tau_1 = \frac{1}{3}, \tau_2 = \frac{1}{5}$.

First, we design the controller for the switched system (4.38) with $g_{1,1}(x) = g_{1,2}(x) = 0$.

We consider the x_1-equation of each subsystem and view x_2 as the input. Choose $x_{2,1}^* = -\beta_{1,1}\xi_{1,1}$ and $x_{2,2}^* = -\beta_{1,2}\xi_{1,2}$ with $\beta_{1,1} = 8, \beta_{1,2} = 32$ and $\xi_{1,1} = \xi_{1,2} = x_1$. One can verify that (4.10) holds with $V_1 = \frac{1}{2}x_1^2$ and $n = 2$.

Based on the virtual control $x_{2,k}^*, k = 1, 2$, the state transformation is different from each other.

Define $\xi_{2,1} = x_2 + 8\xi_{1,1}, \xi_{2,2} = x_2 + 32\xi_{1,2}$. Following the design method in above section, we choose $V_{2,k}(x) = V_{1,k}(x_1) + W_{2,k}(x)$ for subsystems, where $W_{2,k}(x) = \frac{1}{2}(x_2 - x_{2,k}^*)^2$. Then, one can construct controllers for (4.38):

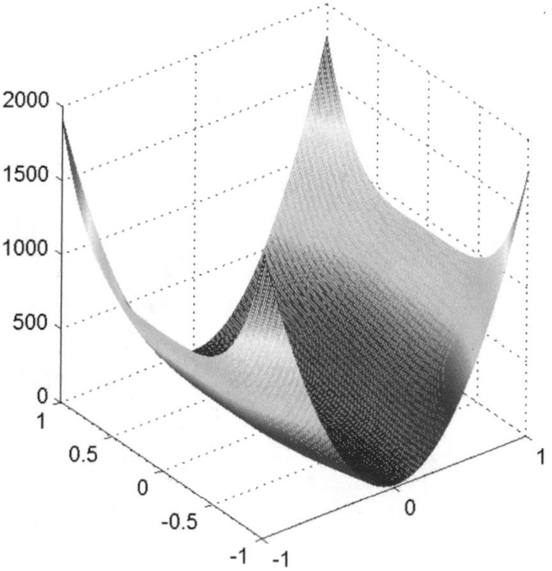

Fig. 4.1 Condition (4.33) with $k = 1$

$$u_k = x_{3,k}^* = -\beta_2(x)\xi_{2,k}^{\frac{1}{q_{3,k}}}, \tag{4.39}$$

where $\beta_{2,1}(x) = \left[\frac{27}{32} + \frac{3}{4}(\frac{2}{\breve{c}_{2,1,1}}2^{-\frac{2}{3}})^{-\frac{1}{3}} + 2^{\frac{2}{3}}\breve{c}_{2,1,1} + L_{2,1}(1+\xi_{2,1}^2) + 1\right]^{\frac{1}{r_{2,1}}}$, $\beta_{2,1}(x) = (\frac{1}{6}(\frac{5}{3})^5 2^{\frac{24}{5}} + \frac{5}{6}(\frac{3}{\breve{c}_{21,k}}2^{-\frac{4}{5}})^{-\frac{1}{5}} + 2^{\frac{4}{5}}\breve{c}_{21,k} + L_{2,2}(1+\xi_{2,2}^2) + 1)^{\frac{1}{r_{2,1}}}$ with $\breve{c}_{2,1,k}$=$32 \times 32^{r_{1k}}$, $L_{2,1} = 0.01 \times (1+g_{11}^2)$, $L_{2,2} = 0.01 \times (1+g_{12}^2)$, such that that the closed-loop system (4.38) with (4.39) when $g_{1,k}(x) = 0$ satisfies $\dot{V}_n(x) + c_k V_n^{\frac{1+\tau_k}{2}}(x) \leq -L_{2,k}\xi_{2,2}^2$, where $c_1 = 2^{-\frac{1+\tau_1}{2}} = 2^{-\frac{2}{3}}$, $c_2 = 2^{-\frac{3}{5}}$.

One can know that (4.33) holds with $w_{1,2}(x) = w_{2,1}(x) = 0.01$, which can be also seen from Figs. 4.1 and 4.2. Then, from Theorem 4.5, the closed-loop system (4.38) with (4.39) under the switching law (4.34) is globally finite-time stable.

The simulation is carried out with the initial states $[-0.2, 0.3]^T$. Figure 4.3 shows the state response of the switched system, which indicates that the closed-loop system is finite-time stable. Figure 4.4 gives the switching signal. Thus, the simulation results well illustrate the effectiveness of the proposed control method.

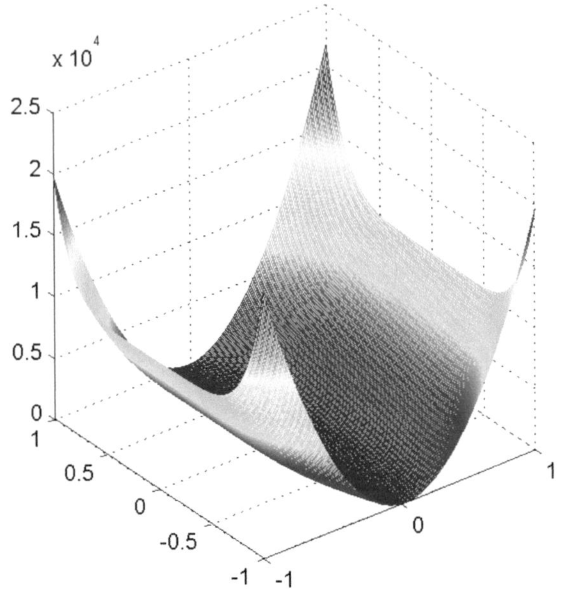

Fig. 4.2 Condition (4.33) with $k = 2$

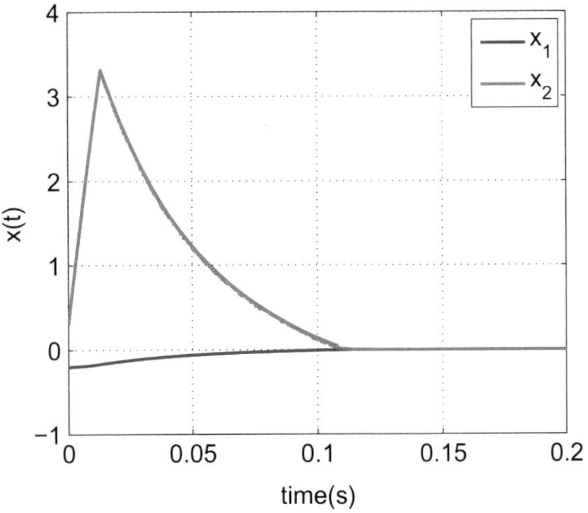

Fig. 4.3 The state response of the switched system

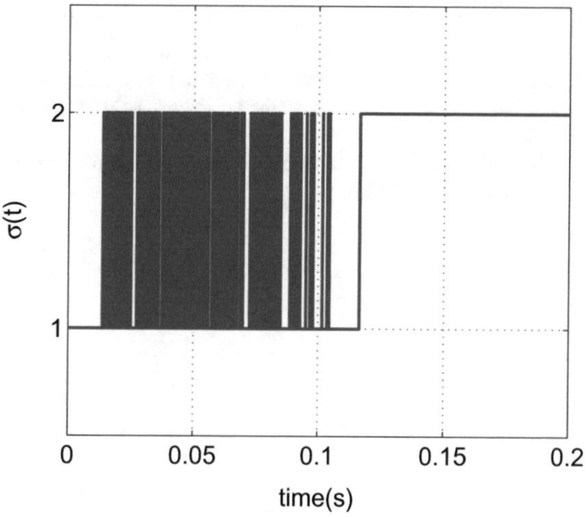

Fig. 4.4 The switching signal for the switched system

4.5 Concluding Remarks

In this chapter, we have investigated the global finite-time stabilization for a class of switched nonlinear systems in non-triangular form where the global stabilization problem of individual subsystems is not assumed to be solvable when applying adding a power integrator technique. Unlike the well-known nonlinear structure such as the high-order nonlinear systems, we have studied the more general nonlinear structure-the non-triangular form. Both new sufficient conditions and controller design method for subsystems were presented. Based on the MLFs method and the adding a power integrator technique, we have designed state feedback controllers of subsystems and constructed a switching law, which guarantees global finite-time stabilization of the corresponding closed-loop system. An example further demonstrated efficiency of the proposed research.

References

1. Long, L., Zhao, J.: H_∞ control of switched nonlinear systems in p-normal form using multiple lyapunov functions. IEEE Trans. Autom. Control **57**(5), 1285–1291 (2012)
2. Long, L., Zhao, J.: Global stabilization of switched nonlinear systems in non-triangular form and its application. J. Frankl. Inst. **365**(2), 1161–1178 (2014)
3. Han, T.-T., Ge, S.S., Lee, H.T.: Adaptive neural control for a class of switched nonlinear systems. Syst. Control Lett. **58**(2), 109–118 (2009)
4. Ma, R., Zhao, J.: Backstepping design for global stabilization of switched nonlinear systems in lower triangular form under arbitrary switchings. Automatica **46**(11), 1819–1823 (2010)
5. Jenq-Lang, W.: Stabilizing controllers design for switched nonlinear systems in strict-feedback form. Automatica **45**(4), 1092–1096 (2009)

6. Long, L., Zhao, J.: Global stabilisation of switched nonlinear systems in p-normal form with mixed odd and even powers. Int. J. Control **84**(10), 1612–1626 (2011)
7. Back, J., Cheong, S.G., Shim, H., Seo, J.H.: Nonsmooth feedback stabilizer for strict-feedback nonlinear systems that may not be linearizable at the origin. Syst. Control Lett. **56**(11–12), 742–752 (2007)
8. Shen, Y., Huang, Y.: Global finite-time stabilisation for a class of nonlinear systems. Int. J. Syst. Sci. **43**(1), 73–78 (2010)

Chapter 5
Adaptive Finite-Time Stabilization of Uncertain Nonlinear Systems via Logic-Based Switchings

In this chapter, the global adaptive finite-time stabilization is investigated by logic-based switching control for a class of uncertain nonlinear systems with the powers of positive odd rational numbers. Parametric uncertainties entering the state equations nonlinearly can be fast time-varying or jumping at unknown time instants, and the control coefficient appearing in the control channel can be unknown. The bounds of the parametric uncertainties and the unknown control coefficient are not required to know a priori. Our proposed controller is a switching-type one, in which a nonlinear controller with two parameters to be tuned is first designed by adding a power integrator, and then a switching mechanism is proposed to tune the parameters online to finite-time stabilize the system. An example is provided to demonstrate the effectiveness of the proposed result.

5.1 Introduction

In previous three chapters, the problem of finite-time stabilization has been investigated for different control systems. However, parametric uncertainties exist in many practical systems. For the control systems with parametric uncertainties, the adaptive control is one of the effective ways [1, 2]. In this chapter, we investigate whether it is possible to apply the logic-based switching technique in [3–5] to achieve adaptive finite-time stabilization of a class of p-normal form nonlinear systems, whose powers of the chained integrators are allowed to be *any* positive odd rational numbers, and whose parametric uncertainties entering the state equations nonlinearly can be fast time-varying or jumping at unknown time instants, and the bounds of which are not required to be known a priori, if yes, under what conditions it can be achieved. To our best knowledge, there have not been results in the literature that provide answers to these questions. For the considered problem, the main difficulties are in twofold. One is that the powers of the chained integrators are allowed to be *any* positive

J. Fu and R. Ma, *Stabilization and H_∞ Control of Switched Dynamic Systems*, Studies in Systems, Decision and Control 310,
https://doi.org/10.1007/978-3-030-54197-2_5

odd rational numbers. The authors in [6] proposed a condition on characterizing the powers that enable the finite-time stabilization. However, the adaptive finite-time stabilization is not considered in [6] for the system with unknown parameters. The other is how to use switching control to copy with adaptive finite-time stabilization of p-form nonlinear systems with these unknown parameters. To overcome these two difficulties, we design a logic-based switching controller, in which a nonlinear controller with two parameters to be tuned is first designed by adding a power integrator [7], and then a modified switching mechanism is proposed to tune the two parameters online to finite-time stabilize the system. Unlike [3–5] where logic switching rules are developed for adaptive asymptotical stabilization, a switching mechanism for adaptive finite-time stabilization is proposed in this paper. Unlike [8], we will use the upper bound function of an integral-type Lyapunov function obtained in constructing the controller to design the switching mechanism. Such a upper bound function will facilitate implementation of the designed switching mechanism. An example is provided to demonstrate the effectiveness of the proposed result.

5.2 System Description

In this chapter, we consider the following uncertain nonlinear systems:

$$\begin{aligned} \dot{x}_i &= x_{i+1}^{r_i} + f_i(x, u, t, \theta(t)), \\ i &= 1, 2, \ldots, n-1, \\ \dot{x}_n &= \delta(t) u^{r_n} + f_n(x, u, t, \theta(t)), \end{aligned} \tag{5.1}$$

where $x = (x_1, \ldots, x_n) \in R^n$ is state, $u \in R$ is control input, $\bar{x}_i = (x_1, \ldots, x_i) \in R^i$, $i = 1, \ldots, n$, $r_i \in Q_{odd}$, $i = 1, \ldots, n$, (Q_{odd} denotes the set of all positive rational numbers whose numerators and denominators are all positive odd integers), f_i, $i = 1, \ldots, n$, are C^1 with $f_i(0, 0, t, \theta(t)) = 0$, parameter vector $\theta(t) \in R^s$ and parameter $\delta(t) \in R$ are unknown, which may be fast time-varying or jumping at unknown time instants, and the bounds of which are not required to be known a priori.

The objective of this chapter is to adaptively finite-time stabilize systems (5.1) whose powers can be *any* positive odd rational numbers and whose parameters may be fast time-varying or jumping at unknown time instants.

Now, we list three technical assumptions below.

Assumption 5.1 ([6]) For given $r_i \in Q_{odd}, i = 1, \ldots, n$, in (5.1), there are constants $\mu_i \in Q_{odd}$, $\mu_i \geq 1$, $i = 0, 1, \ldots, n$, such that

$$0 < \frac{1}{\mu_0} - \frac{r_1}{\mu_1} \leq \frac{1}{\mu_1} - \frac{r_2}{\mu_2} \leq \cdots \leq \frac{1}{\mu_{n-1}} - \frac{r_n}{\mu_n}, \tag{5.2}$$

$$\frac{r_i}{\mu_i} \leq \min\left\{\frac{1}{\mu_0}, \frac{1}{\mu_1}, \frac{1}{\mu_2}, \ldots, \frac{1}{\mu_{i-1}}\right\}, i = 1, \ldots, n. \tag{5.3}$$

Assumption 5.2 ([9]) For $i = 1, \ldots, n$, there exists a known nonlinearly parameterized function $\gamma_i(\bar{x}_i, \theta) \geq 0$, such that

$$|f_i(x, u, \theta)| \leq \gamma_i(\bar{x}_i, \theta) \left(|x_1| + \cdots + |x_i|\right), \forall (x, u, \theta). \tag{5.4}$$

Assumption 5.3 ([3]) The sign of $\delta(t)$ is known and unchanged, and $\bar{\delta} \geq |\delta(t)| > \underline{\delta}$ with unknown constants $\bar{\delta}, \underline{\delta} > 0$.

Remark 5.1 Assumption 5.1 is checkable or tractable as shown in [6]. When $r_i \geq 1$ and $r_i \in Q_{odd}, i = 1, \ldots, n$, Assumption 5.1 naturally holds, which corresponds to the result of [7]. The underactuated unstable mechanical system introduced in [7] is a practical example to satisfy Assumption 5.1. For the case when $\mu_i = 1, i = 0, 1, \ldots, n$, then we get $1 \geq r_1 \geq r_2 \geq \cdots \geq r_n$, and $r_i \in Q_{odd}, i = 1, \ldots, n$, which also satisfies Assumption 5.1, which corresponds to the case in [6]. Assumption 5.2 represents the uncertain nonlinear system is dominated by a lower-triangular one. If the control coefficients, as well as those of the first $n - 1$ subsystems, also depends on system states, our method can be extended to deal with this case by making similar assumptions for these coefficients as $\bar{\delta} \geq |\delta(t, x(t))| > \underline{\delta} > 0$. When $\gamma_i(\bar{x}_i, \theta)$ is known, Assumption 5.2 was used for studying the finite-time stabilization in [9]. Assumption 5.3 is more general than the corresponding one in [6, 9], where there is no uncertain control coefficient. Since the sign of $\delta(t)$ is unchanged, then, in the later design procedure, we assume $\delta(t) > 0$ without loss of generality.

5.3 Adaptive Stabilizer Design

Our proposed controller is a switching-type one, in which a nonlinear controller with two parameters to be tuned is first designed by adding a power integrator, and then a switching mechanism is proposed to tune the parameters online in a piecewise constant way.

5.3.1 Controller Design

Here, we assume that k and ℓ in designed controller are fixed constants. By Assumption 5.1, we choose $\upsilon_0, \ldots, \upsilon_{n-1} \in Q_{odd}$ with $\upsilon_0, \ldots, \upsilon_{n-1} \geq 1$, such that

$$\upsilon_0 + \frac{r_1}{\mu_1} = \upsilon_1 + \frac{r_2}{\mu_2} = \cdots = \upsilon_{n-1} + \frac{r_n}{\mu_n} \triangleq \tau. \tag{5.5}$$

Since the uncertainty θ is bounded, then there is a compact set Ω, such that $\theta \in \Omega$. Thus, like [4], we define

$$p^* = \max_{\theta \in \Omega} \{\|\theta\|\}. \tag{5.6}$$

Note that such a set Ω is actually unknown. Accordingly, p^* in the left-hand side of (5.6) is also an unknown scalar while θ in the right-hand side is an unknown vector. Such an unknown p^* is only associated with the size of the set, and independent on the rate of change of θ. Therefore, it is convenient for the design of our tuning mechanism.

Step 1. For each constant $p \geq 0$, we define

$$\bar{\gamma}_1(x_1, k, p) = \max_{\|\theta\| < p} \{|\gamma_1(x_1, \theta)|\}. \tag{5.7}$$

Note that such a function $\bar{\gamma}_1(x_1, k, p)$ does not explicitly contain k, which is to correspond to the function of the recursive step. We then choose a C^1 function $\tilde{\gamma}_1(x_1, k, p)$, which is strictly increasing with respect to p, i.e., $p_1 > p_2$ implies $\tilde{\gamma}_1(x_1, k, p_1) > \tilde{\gamma}_1(x_1, k, p_2)$, and satisfying

$$\tilde{\gamma}_1(x_1, k, p) > \bar{\gamma}_1(x_1, k, p). \tag{5.8}$$

From (5.6), (5.7) and (5.8), we obtain that

$$\gamma_1(x_1, \theta) \leq \tilde{\gamma}_1(x_1, k, p^*). \tag{5.9}$$

We choose a C^1 function $\hat{\gamma}_1(x_1, k, p)$, which is strictly increasing with respect to p and satisfies

$$\hat{\gamma}_1(x_1, k, p) \geq (1 + |x_1|^{\varrho - \mu_0 \frac{r_1}{\mu_1}})\tilde{\gamma}_1(x_1, k, p), \tag{5.10}$$

where $\varrho \geq 1$. Define a C^1 function $\beta_1(x_1, k, p^*) = (n + \hat{\gamma}_1(x_1, k, p^*))^{\frac{\mu_1}{r_1}}$. Setting p^* as k in $\beta_1(x_1, k, p^*)$, we design

$$x_2^{*r_1}(x_1, k) = -(x_1^{\mu_0})^{\frac{r_1}{\mu_1}} \beta_1^{\frac{r_1}{\mu_1}}(x_1, k, k), \tag{5.11}$$

and a C^1, positive definite and proper function

$$V_1(x_1) = x_1^{\mu_0 v_0 + 1} \triangleq W_1(x_1). \tag{5.12}$$

Then, by (5.1), (5.9), (5.10) and $\mu_0 r_1 / \mu_1 \leq 1$, one has

$$\begin{aligned}
&\dot{V}_1 \\
&\le (x_1^{\mu_0})^{\upsilon_0}(x_2^{r_1} - x_2^{*r_1} + x_2^{*r_1}) + |x_1|^{\mu_0\upsilon_0}|x_1|\gamma_1(x_1,\theta) \\
&\le (x_1^{\mu_0})^{\upsilon_0}(x_2^{r_1} - x_2^{*r_1}) + (x_1^{\mu_0})^{\upsilon_0}x_2^{*r_1} \\
&\quad + |x_1^{\mu_0}|^{(\upsilon_0+\frac{r_1}{\mu_1})}(1 + |x_1|^{\varrho-\mu_0\frac{r_1}{\mu_1}})\tilde{\gamma}_1(x_1,k,p) \\
&\le (x_1^{\mu_0})^{\upsilon_0}(x_2^{r_1} - x_2^{*r_1}) + (x_1^{\mu_0})^{\upsilon_0}x_2^{*r_1} + |x_1^{\mu_0}|^{(\upsilon_0+\frac{r_1}{\mu_1})}\hat{\gamma}_1(x_1,k,p^*) \\
&\le -n(x_1^{\mu_0})^{(\upsilon_0+\frac{r_1}{\mu_1})} + (x_1^{\mu_0})^{\upsilon_0}\left[x_2^{r_1} - x_2^{*r_1}\right] \\
&\quad -(x_1^{\mu_0})^{(\upsilon_0+\frac{r_1}{\mu_1})}\left[\hat{\gamma}_1(x_1,k,k) - \hat{\gamma}_1(x_1,k,p^*)\right].
\end{aligned} \tag{5.13}$$

Remark 5.2 In (5.10), we need to generate $\hat{\gamma}_1$. Here we exemplify how to generate such a function. Define $\varphi(x,s) = \sup_{\|y\|\le s} f(x,y), s \ge 0$. It is obvious that $\varphi(x,s)$ is continuous, increasing with respect to s, and $f(x,y) \le \varphi(x,y)$. Then, one can easily choose a C^1 function $k(x,y)$ which is strictly increasing with respect to y, for example, $k(x,y) \triangleq \frac{1}{2}\left(1 + \varphi^2(x,y)\right) + a\,\|y\|^2$ where $a > 0$, such that $k(x,y)$ satisfies $f(x,y) \le k(x,y)$.

Step i $(i = 2, 3, \ldots, n)$. For $i = 1, \ldots, n$, we define

$$U_i = \sum_{j=1}^{i} \xi_j^{\tau}\left[\hat{\gamma}_j(\bar{\xi}_j,k,k) - \hat{\gamma}_j(\bar{\xi}_j,k,p^*)\right],\ Y_i = \sum_{j=1}^{i} \xi_j^{\tau}.$$

Suppose at step $i-1$, we have had a C^1 function $V_{i-1}(\bar{x}_{i-1})$ and a set of virtual controllers $x_1^*, x_2^*, \ldots, x_i^*$:

$$\begin{array}{ll}
x_1^* = 0, & \xi_1 = x_1^{\mu_0} - x_1^{*\mu_0}, \\
x_2^{*\mu_1} = -\beta_1(x_1,k)\xi_1, & \xi_2 = x_2^{\mu_1} - x_2^{*\mu_1}, \\
\vdots & \vdots \\
x_i^{*\mu_{i-1}} = -\beta_{i-1}(\bar{x}_{i-1},k)\xi_{i-1}, & \xi_i = x_i^{\mu_{i-1}} - x_i^{*\mu_{i-1}},
\end{array} \tag{5.14}$$

with C^1 functions $\beta_j^{r_j/\mu_j}(\bar{x}_j,k)$, $1 \le j \le i-1$, such that

$$\dot{V}_{i-1} \le -(n-i+2)Y_{i-1} - U_{i-1} + \xi_{i-1}^{\upsilon_{k-2}}(x_i^{r_{i-1}} - x_i^{*r_{i-1}}). \tag{5.15}$$

Next, we show that (5.15) holds at Step i. To prove this claim, we consider a Lyapunov function candidate defined by

$$V_i = V_{i-1} + W_i = V_{i-1} + \int_{x_i^*}^{x_i} (s^{\mu_{i-1}} - x_i^{*\mu_{i-1}})^{\upsilon_{i-1}} ds. \tag{5.16}$$

One has that $V_i(\bar{x}_i)$ is C^1, positive definite and proper with properties:

$$\frac{\partial W_i}{\partial x_j} = -\int_{x_i^*}^{x_i} \upsilon_{i-1}(s^{\mu_{i-1}} - x_i^{*\upsilon_{i-1}})^{\upsilon_{i-1}-1} ds \frac{\partial (x_i^{*\mu_{i-1}})}{\partial x_j}, \tag{5.17}$$

$$\frac{\partial W_i}{\partial x_i} = (x_i^{\mu_{i-1}} - x_i^{*\mu_{i-1}})^{\upsilon_{i-1}} = \xi_i^{\upsilon_{i-1}}, \ j = 1, \ldots, i-1. \tag{5.18}$$

which can be obtained by the similar way in [6].

We now can continue constructing the controller. Using (5.1) and (5.15)–(5.18), it can be deduced that

$$\begin{aligned} \dot{V}_i \le & -(n-i+2)Y_{i-1} - U_{i-1} + \xi_{i-1}^{\upsilon_{i-2}}(x_i^{r_{i-1}} - x_i^{*r_{i-1}}) \\ & + \xi_i^{\upsilon_{i-1}}\left[x_{i+1}^{r_i} + f_i\right] + \sum_{j=1}^{i-1} \frac{\partial W_i}{\partial x_j}\dot{x}_j. \end{aligned} \tag{5.19}$$

Next, to estimate the last three terms on the right-hand side of (5.19), we give the following proposition for system (5.1), which can be obtained by the similar way in [10].

Proposition 5.3 *There are constants $c_{i1}(k) > 0$ and C^1 functions $c_{ij}(\bar{x}_i, k, \theta) > 0$, $j = 2, 3$, such that*

$$\left|\xi_{i-1}^{\upsilon_{i-2}}(x_i^{r_{i-1}} - x_i^{*r_{i-1}})\right| \le \frac{1}{3}Y_{i-1} + c_{i1}(k)\xi_i^{\upsilon_{i-1}+r_i/\mu_i}, \tag{5.20}$$

$$\left|\xi_i^{\upsilon_{i-1}} f_i\right| \le \frac{1}{3}Y_{i-1} + c_{i2}(\bar{x}_i, k, \theta)\xi_i^{\upsilon_{i-1}+r_i/\mu_i}, \tag{5.21}$$

$$\sum_{j=1}^{i-1} \frac{\partial W_i}{\partial x_j}\dot{x}_j \le \frac{1}{3}Y_{i-1} + c_{i3}(\bar{x}_i, k, \theta)\xi_i^{\upsilon_{i-1}+r_i/\mu_i}. \tag{5.22}$$

We choose a C^1 function $\hat{\gamma}_i(\bar{x}_i, k, p)$, which is strictly increasing with respect to p, such that

$$\hat{\gamma}_i(\bar{x}_i, k, p) \ge \bar{\gamma}_i(\bar{x}_i, k, p), \tag{5.23}$$

where $\bar{\gamma}_i(\bar{x}_i, k, p) = \max_{\|\theta\| \le p} \{c_{i2}(\bar{x}_i, k, \theta) + c_{i3}(\bar{x}_i, k, \theta)\} + c_{i1}(k)$. By (5.6), we further define a C^1 function

$$\beta_i(\bar{x}_i, k, p^*) = (\hat{\gamma}_i(\bar{\xi}_i, k, p^*) + n - i + 1)^{\mu_i/r_i}. \tag{5.24}$$

Then, noticing (5.24) and similarly setting p^* as k, we design

$$x_{i+1}^{*r_i}(\bar{x}_i, k) = -\beta_i^{r_i/\mu_i}(\bar{x}_i, k, k)\xi_i^{r_i/\mu_i}, \tag{5.25}$$

which together with (5.5), (5.19) and Proposition 5.3 gives

$$\begin{aligned}
&\dot{V}_i \\
&\le -(n-i+1)Y_{i-1} - U_{i-1} + \big[c_{i1}(k) + c_{i2}(\bar{x}_i, k, \theta) + c_{i3}(\bar{x}_i, k, \theta) \\
&\quad -\beta_i^{r_i/\mu_i}(\bar{x}_i, k, k)\big]\xi_i^{\tau} + \xi_i^{\upsilon_{i-1}}(x_{i+1}^{r_i} - x_{i+1}^{*r_i}) \\
&\le -(n-i+1)Y_{i-1} - U_{i-1} + \big[\hat{\gamma}_i(\bar{\xi}_i, k, p^*) - \hat{\gamma}_i(\bar{\xi}_i, k, k) \\
&\quad -(n-i+1)\big]\xi_i^{\tau} + \xi_i^{\upsilon_{i-1}}(x_{i+1}^{r_i} - x_{i+1}^{*r_i}) \\
&= -(n-i+1)Y_i - U_i + \xi_i^{\upsilon_{i-1}}(x_{i+1}^{r_i} - x_{i+1}^{*r_i}).
\end{aligned} \tag{5.26}$$

At Step n, we construct the controller as

$$u^{r_n}(x, k, \ell) = x_{n+1}^{*r_n} = -\frac{1}{\ell}\beta_n^{\frac{r_n}{\mu_n}}(x, k, k)\xi_n^{\frac{r_n}{\mu_n}}, \tag{5.27}$$

such that

$$\begin{aligned}
\dot{V}_n &\le -Y_{n-1} - U_{n-1} + \xi_n^{\tau}\hat{\gamma}_n(x, k, p^*) + \xi_n^{\upsilon_n}(\delta(t)u^{r_n} - \ell x_{n+1}^{*r_n}) + \xi_n^{\upsilon_n}\ell x_{n+1}^{*r_n} \\
&\le -Y_n - U_n - \frac{\delta(t) - \ell}{\ell}\big[1 + \hat{\gamma}_n(x, k, k)\big]\xi_n^{\tau}.
\end{aligned} \tag{5.28}$$

By Assumption 5.1, choose a constant α, such that, $\forall i = 1, \ldots, n$,

$$\frac{\upsilon_{i-1} + \frac{r_i}{\mu_i}}{\upsilon_{i-1} + \frac{1}{\mu_{i-1}}} < \alpha < 1. \tag{5.29}$$

Moreover, for $j = 1, 2, \ldots, n$, we have

$$W_j^{\alpha} \le 2|\xi_j|^{\upsilon_{i-1} + \frac{r_i}{\mu_i}} = 2|\xi_j|^{\tau}. \tag{5.30}$$

Then, $\forall x \in R^n$, we obtain

$$V_n^{\alpha} = \big[\sum_{j=1}^{n} W_j(\bar{x}_j)\big]^{\alpha} \le \sum_{j=1}^{n} W_j^{\alpha}(\bar{x}_j) \le 2\sum_{j=1}^{n}\xi_j^{\tau}. \tag{5.31}$$

If we take $c \le \frac{1}{2}$, we obtain

$$\dot{V}_n \le -cV_n^{\alpha} - U_n - \frac{\delta(t) - \ell}{\ell}\big[1 + \hat{\gamma}_n(\bar{\xi}_n, k, k)\big]\xi_n^{\tau}. \tag{5.32}$$

5.3.2 Switching Mechanism

Now, we propose a switching mechanism to tune k and ℓ online in a piecewise constant way. Define

$$V_{max}(x) = [2(\xi_1^\tau + \cdots + \xi_n^\tau)]^{\frac{1}{\alpha}}. \tag{5.33}$$

and two sequences

$$H_1 = \{h_1(i) > 0, i = 1, 2, \ldots, \infty\} \tag{5.34}$$

and

$$H_2 = \{h_2(i) > 0, i = 1, 2, \ldots, \infty\}, \tag{5.35}$$

where the function h_1 is strictly increasing with $\lim_{i\to\infty} h_1(i) = \infty$, and h_2 is strictly decreasing with $\lim_{i\to\infty} h_2(i) = 0$.

Initialization:

- choose a constant $\Delta > 0$;
- let $k = h_1(1)$, $\ell = h_2(1)$ and $t_s = 0$[1];

Switching logic:

- at each time $t > t_s$, when $\|x(t)\| > 0$, if

$$V_n^{1-\alpha}(t) > \max\left\{V_{\max}^{1-\alpha}(t_s) + (1-\alpha)[\Delta - c(t-t_s)], 0\right\}, \tag{5.36}$$

- then, switch k to its next (and larger) element in H_1, switch ℓ to its next (and smaller) element in H_2, and reset $t_s = t$.
- The aforementioned process is then repeated.

Here, $V_n(t)$ is used to denote $V_n(x(t))$. For clarity, we depict our switching mechanism in Fig. 5.1.

Our designed adaptive controller can be viewed as a switching type controller [3–5]. We only test the monitoring signal generator (5.36) rather than calculating the single integral signal of system state in [3–5]. However, by using these switching rules in [3–5], only asymptotical stability of the closed-loop control systems can be guaranteed with these developed adaptive switching controllers. For the design of adaptive finite-time stabilizer in [8], an integral-type Lyapunov function is constructed and then used to design switching logical. Unlike [8], we use $V_{\max}$ in switching mechanism. According to the switching logic, k and ℓ are jumping at switching times t_s, but $x(t)$ is everywhere continuous. Thus, $V_{\max}(t_s)$ is updated by the value of k just after switching. Moreover, it is easy to choose H_1 and H_2, for example, as $H_1 = \{h_1(i+1) = 0.02 + h(i) : h(1) = 0.01, i = 1, \ldots, \infty\}$ and

---

[1] t_s memories the latest switching time, although at the beginning t_s is set equal to 0.

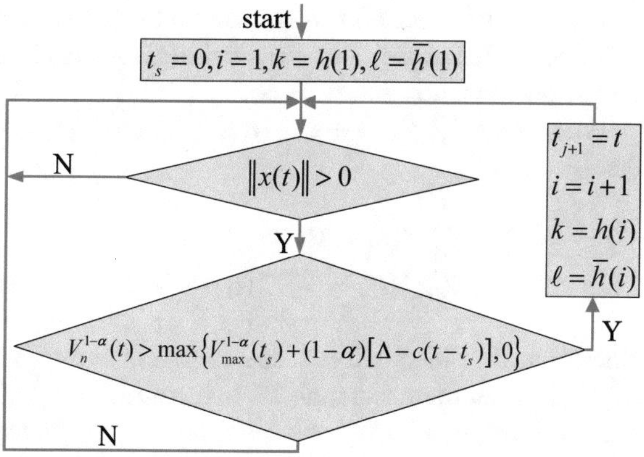

Fig. 5.1 The switching mechanism

$H_2 = \{h_2(i+1) = 0.98h_2(i) : h_2(1) = 5, i = 1, \ldots, \infty\}$, and different H_1 and H_2 yield different switches, which will further lead to different trajectories.

5.4 Main Results

Now, we summarize our main results as follows.

Theorem 5.4 *Consider system (5.1) under Assumptions 5.1–5.3. Suppose the adaptive finite-time stabilizer (5.27) with switching mechanism developed in Sect. 5.3.2 is applied to system (5.1). Then, for any initial conditions, the solution $x(t)$ to the closed-loop system (5.1) exists and is bounded on $[0, \infty)$, and moreover, the state $x(t)$ converges to 0 in a finite time.*

Proof For any fixed k and ℓ, the closed-loop solution exists till the state x escapes to infinity or else the time t equals to infinity. During any two successive switchings, the positive definite function $V_{\max}(t)$ (with fixed k) has increased $\Delta - c(t - t_s)$, and then the closed-loop solution must exist and is bounded. So, the finite time escape phenomenon cannot occur. Let the maximum interval of the solution of the closed-loop system be $\left[0, t_f\right)$.

Next, we will show that if there is only a finite number of switchings, then $t_f = \infty$, and $x(t) \to 0$ as $t \to T^\diamond$, $x(t) \equiv 0$ as $t \geq T^\diamond$, where $T^\diamond$ is a finite number.

Without loss of generality, we assume that the last switching occurs at time t_1. According to the switching logic, when $t > t_1$, we must have:

$$V_n^{1-\alpha}(t) \leq \max\left\{V_{\max}^{1-\alpha}(t_1) + (1-\alpha)[\Delta - c(t - t_1)], 0\right\}. \tag{5.37}$$

Otherwise, a further switching will occur.

If (5.37) holds for all $t > t_1$, then no further switching occurs for $t > t_1$, and V_n is bounded. Therefore, both $x(t)$ and $u(t)$ are also bounded. Thus, there is no finite escape phenomenon and $t_f = \infty$.

First, when $V_{\max}^{1-\alpha}(t_1) + (1-\alpha)[\Delta - c(t-t_1)] > 0$ and $\|x(t_1)\| > 0$, we obtain that

$$\begin{aligned} & V_n^{1-\alpha}(t) - V_n^{1-\alpha}(t_1) \\ & \leq V_{\max}^{1-\alpha}(t_1) - V_n^{1-\alpha}(t_1) + (1-\alpha)[\Delta - c(t-t_1)], \forall t > t_1, \end{aligned} \tag{5.38}$$

where the first inequality is due to (5.31) and (5.33). Since the boundedness of $V_{\max}^{1-\alpha}(t_1) + \Delta$, there is a time $T^* > t_1$, such that $V_{\max}^{1-\alpha}(t_1) + \Delta - c(T^* - t_1)$ equals to zero. Then, by (5.38), we have $V_n(t) \equiv 0$ and hence $x(t) \equiv 0$ for $t \geq T^* > t_1$. In addition, if $\|x(t_1)\| = 0$, one has that $x(t) \equiv 0$ for $t \geq t_1$.

Second, when $V_{\max}^{1-\alpha}(t_1) + (1-\alpha)[\Delta - c(t-t_1)] \leq 0$ for a finite time $T^{\times} > t_1$, we have $V_{\max}^{1-\alpha}(t) = 0$ and hence $x(t) \equiv 0, \forall t \geq T^{\times}$.

In summary, $x(t) \equiv 0$ for $t \geq \max\{T^*, T^{\times}\} \geq t_1$.

Now, we prove that there is only a finite number of switchings. To seek a contradiction, suppose on the contrary that an infinite number of switchings occur. Due to the boundedness of θ and $\delta(t)$ and the properties of H_1 and H_2, there is a switching time t_2 such that $k(t_2) \geq \|\theta\|$ and $|\delta(t)| \geq \ell(t_2)$. Let t_3 be the successive switching time after t_2. So, there is no switching occurring during (t_2, t_3). Then, if $\|x(t)\| > 0$, we must obtain that

$$V_n^{1-\alpha}(t_3^-) > \max\{V_{\max}^{1-\alpha}(t_2) + (1-\alpha)[\Delta - c(t_3 - t_2)], 0\}. \tag{5.39}$$

Since $\hat{\gamma}_l(\bar{\xi}_l, k, p)$ are strictly increasing with respect to p, then we have

$$\hat{\gamma}_l(\bar{\xi}_l(t), k(t_2), k(t_2)) \geq \hat{\gamma}_l(\bar{\xi}_l(t), k(t_2), p^*), 1 \leq l \leq n. \tag{5.40}$$

When $t \in (t_2, t_3)$, by (5.32) and (5.40), we have

$$\begin{aligned} \dot{V}_n(t) \leq & -\sum_{l=1}^{n} \xi_l^{\tau}[\hat{\gamma}_l(\bar{\xi}_l(t), k(t_2), k(t_2)) - \hat{\gamma}_l(\bar{\xi}_l(t), k(t_2), p^*)] \\ & -[\delta(t) - \ell(t_2)]/\ell(t_2)(1 + \hat{\gamma}_n(x, k, k))\xi_n^{\tau} - cV_n^{\alpha}(t). \end{aligned} \tag{5.41}$$

Then, one has

$$\begin{aligned} V_n^{1-\alpha}(t) & \leq V_n^{1-\alpha}(t_2) - (1-\alpha)c(t-t_2) \\ & \leq V_{\max}^{1-\alpha}(t_2) + (1-\alpha)[\Delta - c(t-t_2)]. \end{aligned} \tag{5.42}$$

Thus, $V_n^{1-\alpha}(t)$ is strictly decreasing for $t \geq t_2$. Therefore, it implies that (5.39) does not hold: a contradiction yields. Therefore, the switching stops in a finite time. Hence, the theorem is proved.

Remark 5.5 According to our switching mechanism, only finite switchings occur, say N times. The energy function V can increase or decrease at the first $N-1$ switching instants. However, *right after* the last switching happens, the energy function V starts decreasing to zero.

Remark 5.6 [8] is the first work to address the global finite time stabilization problem for nonlinear systems with multiple unknown control directions. In [8], the power is not greater than 1 and the nonlinear function f_i is bounded above by a *known* function. However, in our paper the power can be greater than 1, which makes system (1) not feedback linearizable and its Jacobian linearization uncontrollable. Moreover, the nonlinear function f_i in system (1) is bounded above by a known nonlinearly parameterized function $\gamma_i(\bar{x}_i, \theta)$ with respect to unknown parameter θ.

Remark 5.7 Compared to [11], the features of this paper are that our method can cope with the system containing fast time-varying parameters or jumping parameters at unknown time instants by using logic-based switchings, and can cover a class of p-normal form system that may not be linearizable at the origin (i.e., when the powers $r_i < 1$). This work can be seen as an extension of [11] when the parametric uncertainty is constant or slow time-varying.

5.5 An Illustrative Example

Consider the following uncertain nonlinear system:

$$\begin{aligned} \dot{x}_1 &= x_2^{\frac{1}{3}} + x_1 e^{\theta(t)x_1}, \\ \dot{x}_2 &= \delta(t) u^{\frac{1}{5}}, \end{aligned} \tag{5.43}$$

where $\theta(t)$ and $\delta(t) > 0$ are uncertainties, the bounds of which are not required to be known a priori.

When $\theta = \delta = 1$, system (5.43) becomes the one in [6] where the finite-time stabilization is achieved. However, the method in [6] is not applied to system (5.43) due to the unknown boundedness of θ and δ. The methods in [4, 5] are not applicable for system (5.43) since the powers of the chained integrators are with positive odd rational numbers. However, our proposed method can achieve the adaptive finite-time stabilization.

With $\mu_0 = \mu_1 = \mu_2 = 1$ and $v_0 = 1$, $v_1 = v_0 + r_1 - r_2 = \frac{17}{15}$, we choose $V_1(x_1) = \frac{x_1^2}{2}$ and then obtain

$$u = -\frac{1}{\ell^5}(x_2 - x_2^*)[1 + \left(\frac{9}{4}\right)^3 \sqrt[3]{4} + c_3(x, k, k)]^5, \tag{5.44}$$

where $x_2^* = -x_1\big[2 + (1 + 0.15x_1^2)\Delta\big]^3 \triangleq -x_1\Xi^3$, $c_3 = 17/15[(14/15 + 4/45x_1^2)\sqrt[3]{A^4} + (14/15 + 1/15x_1^2)B]$, $\Delta = e^{k(1+x_1^2)/2}$, $A = 2\nabla[1 + (1 + 0.15x_1^2)\Delta]$, $B = \sqrt[3]{4}\nabla$ and $\nabla = -\Xi^3 - 3x_1^2\Xi^2\big[0.3\Delta + (1 + 0.15x_1^2)ke^{k(1+x_1^2)/2}\big]$.

Noting that the parameters k and ℓ are adaptively tuned by switching mechanism in Sect. 5.3 with

$$H_1 = \{h_1(i+1) = 0.02 + h(i) : h(1) = 0.01, i = 1, \dots, \infty\}, \tag{5.45}$$

and

$$H_2 = \{h_2(i+1) = 0.98h_2(i) : h_2(1) = 5, i = 1, \dots, \infty\}, \tag{5.46}$$

$V_{max}(x) = [2\xi_1^{4/3} + 2\xi_2^{4/3}]^{1/\alpha}$, $\alpha = \frac{4}{5}$, $\Delta = 50$ and $c = \frac{1}{2}$.

The simulation is carried out with the initial state $x(0) = (-1, -6)^T$, and the "unknown" uncertainties are chosen as

$$\delta(t) = \begin{cases} -2.5t + 2.2, & t \in [0, 1/7)\,; \\ 2.5t + 0.8, & t \in [1/7, 1/3)\,; \\ 1 + 0.7\,|\sin(5t)|\,, & t \in [1/3, 1.5)\,; \end{cases} \tag{5.47}$$

and

$$\theta(t) = \begin{cases} 2 - 2t - |\cos(6t)| & t \in [0, 1/4)\,; \\ 1 + 3\,|\sin(5t/2)|\,, & t \in [1/4, 4/3)\,; \\ 1 + |\cos(6t)|\,, & t \in [4/3, 1.5)\,, \end{cases} \tag{5.48}$$

which are depicted in Fig. 5.2.

Figure 5.3 shows both the state trajectories and the control input of the closed-loop system (5.43). It can be clearly observed from Fig. 5.3 that all of the state variables converge to zero in a finite time. The tuning parameters are illustrated in Fig. 5.4, which shows that there is only a finite number of switchings.

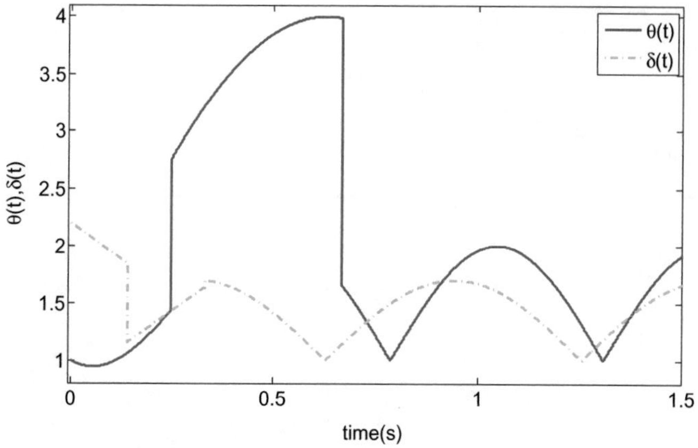

Fig. 5.2 The "unknown" uncertainties $\theta(t)$ and $\delta(t)$

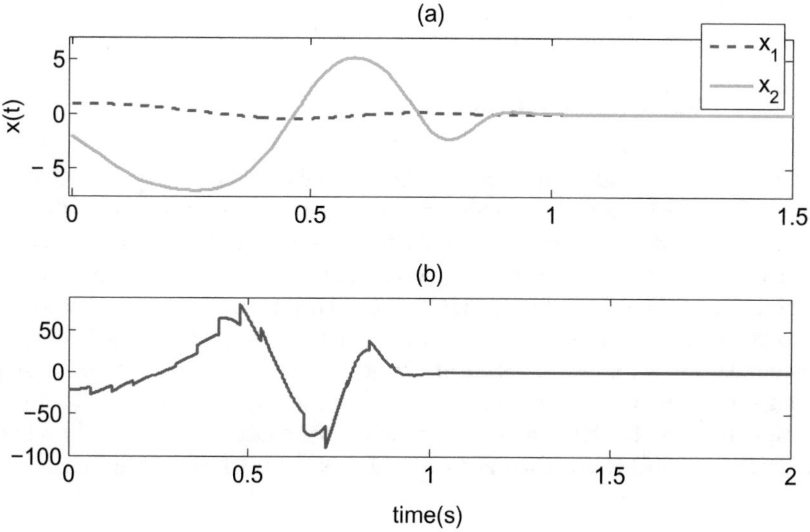

Fig. 5.3 The trajectories of state x and control input u

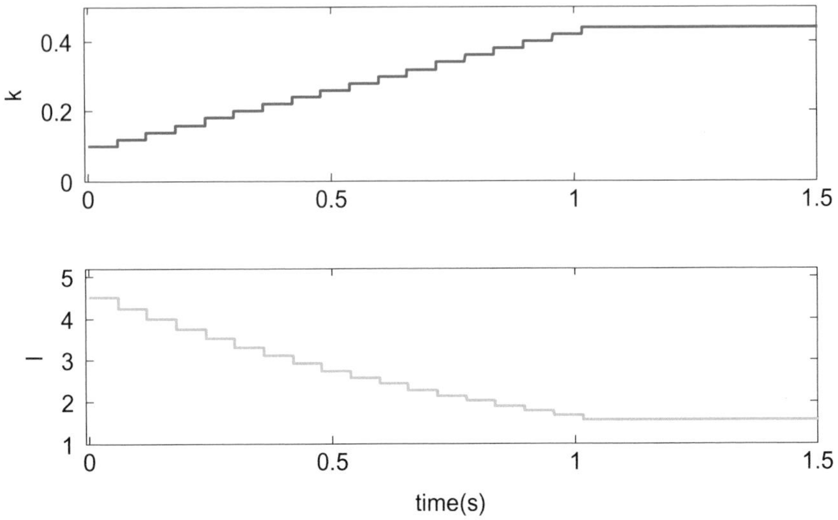

Fig. 5.4 The tuning parameters

5.6 Concluding Remarks

This chapter has studied the global finite-time stabilization by logic-based switching control for a class of uncertain nonlinear systems with the powers of positive odd rational numbers. In this chapter, the parametric uncertainties entering the state equations nonlinearly can be fast time-varying or jumping at unknown time instants, and the control coefficient appearing in the control channel can be unknown. The bounds of the parametric uncertainties and the unknown control coefficient are not required to know a priori. A systematic design method have been proposed to design a logic-based switching controller to finite-time stabilize the systems. Global finite-time stability of the closed-loop system has been proved. This theoretical result can provide a possibly alternative finite-time stabilization method for engineers.

References

1. Jun, F., Chai, T., Chunyi, S., Xie, W.: Adaptive output tracking control of a class of nonlinear systems. Control Eng. China **22**(4), 731–736 (2015)
2. Wang, X., Zhao, J.: Logic-based reset adaptation design for improving transient performance of nonlinear systems. IEEE/CAA J. Autom. Sin. **2**(4), 440–448 (2015)
3. Ma, R., Liu, Y., Zhao, S., Wang, M., Zong, G.: Nonlinear adaptive control for power integrator triangular systems by switching linear controllers. Int. J. Robust Nonlinear Control **25**(14), 2443–2460 (2015)
4. Ye, X.: Global adaptive control of nonlinearly parametrized systems. IEEE Trans. Autom. Control **48**(1), 169–173 (2003)

5. Ye, X.: Nonlinear adaptive control by switching linear controllers. Syst. Control Lett. **61**(4), 617–621 (2012)
6. Back, J., Cheong, S.G., Shim, H., Seo, J.H.: Nonsmooth feedback stabilizer for strict-feedback nonlinear systems that may not be linearizable at the origin. Syst. Control Lett. **56**(11–12), 742–752 (2007)
7. Qian, C., Lin, W.: Non-lipschitz continuous stabilizers for nonlinear systems with uncontrollable unstable linearization. Syst. Control Lett. **42**(3), 185–200 (2001)
8. Jian, W., Chen, W., Li, J.: Global finite-time adaptive stabilization for nonlinear systems with multiple unknown control directions. Automatica **69**, 298–307 (2016)
9. Huang, X., Lin, W., Yang, B.: Global finite-time stabilization of a class of uncertain nonlinear systems. Automatica **41**(5), 881–888 (2005)
10. Jun, F., Ma, R., Chai, T.: Global finite-time stabilization of a class of switched nonlinear systems with the powers of positive odd rational numbers. Automatica **54**(4), 360–373 (2015)
11. Hong, Y., Wang, J., Cheng, D.: Adaptive finite-time control of nonlinear systems with parametric uncertainty. IEEE Trans. Autom. Control **51**(5), 858–862 (2006)

Chapter 6
Dwell-Time-Based Standard H_∞ Control of Switched Systems Without Requiring Internal Stability of Subsystems

This chapter investigates standard H_∞ control of switched systems via dwell time switchings without posing any internal stability requirements on subsystems of the switched systems. First, a sufficient condition is formed by specifying lower and upper bounds of the dwell time, constraining upper bound of derivative of Lyapunov function of the active subsystem, and forcing the Lyapunov function values of the overall switched system to decrease at switching times to achieve standard H_∞ control of unforced switched linear systems. Then, in the same framework of the dwell time, sufficient conditions are given for that of the corresponding forced switched linear systems by further designing state feedback controllers. Finally, numerical examples are provided to demonstrate the effectiveness of the proposed results.

6.1 Introduction

The objective of H_∞ control is to design a control law for a system such that the closed-loop system is internally stable, meanwhile guaranteeing L_2-gain from external disturbances to its controlled output less than or equal to a certain prespecified constant. H_∞ control has been extensively studied in the literature (see, for example, [1, 2], and references therein), which can cope with many robustness problems such as sensitivity minimization [3] and stabilization of uncertain systems [4, 5]. In addition, switched systems have been attracted much attention because of their importance from both theoretical and practical points of view [6–17]. Many practical applications can be formulated as H_∞ control of switched systems such as networked control systems [18] and target tracking systems [19]. Thus, this paper focuses on H_∞ control of switched systems.

J. Fu and R. Ma, *Stabilization and H_∞ Control of Switched Dynamic Systems*, Studies in Systems, Decision and Control 310,
https://doi.org/10.1007/978-3-030-54197-2_6

For H_∞ control of switched systems, there are mainly three main methods: arbitrary switchings, state-dependent switchings, and dwell-time-based methods [20–26]. Xu and Teo in [25] proposed a switched Lyapunov function for a class of switched impulsive systems to achieve H_∞ property under arbitrary switchings. The authors of [20–23, 26] developed state-dependent switching strategies for H_∞ control problems of those switched systems with their own spectacular features. For the dwell-time-based method, the references [20, 24, 27–33] exploited the feature of the dwell time technique: a dwell time of active subsystem can subside possible large state transients.

To the considered H_∞ control problem in this paper, the most relevant papers are [20, 24, 27–33]. References [20, 27] both investigated H_∞ control problems for switched linear systems in continuous time and discrete time, respectively. However, these results are under the assumption that all subsystems are either stable or stabilizable. References [24, 29–33] require that all subsystems are stable or at least one is stable, and only guarantee a weaker weighted (by a decreasing exponential function) disturbance attenuation level than a standard L_2-gain property. When the power of the decreasing exponential function (e.g., λ in [32]) approaches zero, the weighted disturbance attenuation level goes to the attenuation level of standard L_2-gain property. *However, the power of the exponential function (i.e., the above λ) approaching zero results that dwell time of switched systems goes to infinity.* This fact requires that there must be at least one *stable* subsystem of the switched system to achieve the internal stabilization. Although the work by Geromel and Colaneri [28] ensures the standard L_2-gain performance for a class of switched linear systems, the authors pose a conservative assumption that all subsystems of the switched system are stable. It is known that stabilization of switched continuous-time systems with all modes unstable has been achieved by the dwell time switching technique in [34–36]. Thus, questions naturally arise: is it possible, in the framework of the dwell time technique, to achieve H_∞ control of unforced or forced switched systems with the standard L_2-gain performance *without posing any internal stability requirements on subsystems of the switched systems*? If possible, under what conditions can we come up with a switching control to achieve this goal and how? To our best knowledge, in the literature there have not been results which provide answers to these questions. This is the motivation of the present paper.

To address the problems above, first a type of multiple time-varying Lyapunov functions is utilized to achieve standard L_2-gain performance for unforced switched linear systems meanwhile ensuring their internal stability under a sufficient condition, which specifies lower and upper bounds of the dwell time, constrains upper bound of derivative of Lyapunov function of the active subsystem, and forces the Lyapunov function values of the overall switched system decrease at switching times of the overall unforced switched linear systems. Then sufficient conditions are given in the same framework of the dwell time for that of the corresponding forced switched linear systems by further designing state feedback controllers. The main contributions of this manuscript are summarized as follows:

1. A type of multiple time-varying Lyapunov functions is utilized to achieve standard L_2-gain performance for *unforced* switched linear systems meanwhile ensuring their internal stability of the overall switched system without requiring internal stability of subsystems.
2. Sufficient conditions are given in the framework of the dwell time for that of the corresponding *forced* switched linear systems by designing state feedback controllers.
3. The obtained controller gains are time-independent, which can bring a potential advantage in terms of implementation.

This paper is organized as follows. Section 6.2 gives the problem statement. The H_∞ control for unforced switched linear systems is presented in Sect. 6.3. Section 6.4 presents the state feedback H_∞ controllers for forced switched linear systems. Section 6.6 gives three illustrative examples to verify the effectiveness of the proposed results. Section 6.7 concludes this chapter with some brief remarks.

6.2 Problem Statement

In this chapter, we consider the following class of switched linear systems:

$$\begin{aligned} \dot{x} &= A_{\sigma(t)}x + B_{\sigma(t)}u_{\sigma(t)} + E_{\sigma(t)}w, \\ y &= C_{\sigma(t)}x + D_{\sigma(t)}w + F_{\sigma(t)}u_{\sigma(t)}, \end{aligned} \tag{6.1}$$

where $x(t) \in R^n$ is the state, $y(t) \in R^{n_1}$ is the controlled output, $w(t) \in R^{n_2}$ is the disturbance input with $w(t) \in L_2[0, \infty)$, $\sigma(t) : R_+ = [0, \infty) \to M = \{1, 2, \ldots, m\}$ is the switching law which is assumed to be a piecewise continuous (from the right) function of time, with m being the number of subsystems, u_k, $\forall k \in M$, is the control input of the kth subsystem, A_k, B_k, E_k, C_k, and D_k, $\forall k \in M$, are known matrices of the appropriate dimensions. Define $\mathcal{T}_{[\underline{\tau}, \bar{\tau}]}$ as the set of all $\sigma(t)$, where the time interval between successive discontinuities satisfies $t_{i+1} - t_i \in [\underline{\tau}, \bar{\tau}]$, $i = 0, 1, \ldots$, subject to $\bar{\tau} \geq \underline{\tau} > 0$. $N_\sigma(s, t)$ denotes the number of switchings of $\sigma(t)$ over the interval (s, t). We assume that the state of system (6.1) does not jump at switching times, i.e., $x(t)$ is everywhere continuous [26].

The control objective of this chapter is to design a continuous feedback controller u_k for each subsystem $k \in M$, and a switching law $\sigma(t) \in \mathcal{T}_{[\underline{\tau}, \bar{\tau})}$, such that:

(i) the origin of the closed-loop switched system (6.1) is globally asymptotically stable (GAS) when $w \equiv 0$;
(ii) the overall L_2-gain from w to y of the closed-loop switched system (6.1) with the initial state $x(t_0) = 0$ is less than or equal to γ, where $\gamma > 0$ is a constant, i.e.,

$$\int_{t_0}^{\infty} \|y(s)\|^2 ds \leq \gamma^2 \int_{t_0}^{\infty} \|w(s)\|^2 ds. \tag{6.2}$$

Note that if at least one subsystem of system (6.1) satisfies the above (i) and (ii), such a standard H_∞ control problem of system (6.1) is trivial, since if this is the case, we can simply make this subsystem active during whole operating time interval such that the standard H_∞ control problem of system (6.1) is solvable. Therefore, none of the subsystems is assumed to satisfy the above (i) and (ii). This is also a feature of this chapter.

6.3 H_∞ Control for Unforced Switched Linear Systems

In this section, we first present the H_∞ control criterion for the switched system (6.1) with $u(t) \equiv 0$, i.e., for the unforced switched systems:

$$\begin{aligned} \dot{x} &= A_{\sigma(t)}x + E_{\sigma(t)}w, \\ y &= C_{\sigma(t)}x + D_{\sigma(t)}w. \end{aligned} \tag{6.3}$$

The main result for system (6.3) is given below.

Theorem 6.1 *For given constants $\gamma > 0$, $\lambda > 0$, $\Delta_2 \geq \Delta_1 > 0$, and $\mu_{kl} > 1$, $k \in M$, $l = 1, 2$, if there exist matrices $P_{k1} > 0$ and $P_{k2} > 0$, $k \in M$, such that*

$$-\mu_{k_2 2}P_{k_1 1} + P_{k_2 2} < 0, \forall k_1, k_2 \in M,\ k_1 \neq k_2, \tag{6.4}$$

$$\Theta_{klq} = \begin{bmatrix} \theta_{klq}^{(1,1)} - \lambda P_{kl} & P_{kl}E_k & C_k^T \\ E_k^T P_{kl} & -\gamma^2 I & D_k^T \\ C_k & D_k & -\frac{1}{\mu_{k1}\mu_{k2}} I \end{bmatrix} < 0, \tag{6.5}$$

$$ln\tilde{\mu} + \lambda\Delta_2 < 0, \tag{6.6}$$

where

$$\theta_{klq}^{(1,1)} = \vartheta_k P_{kl} + A_k^T P_{kl} + P_{kl}A_k + \frac{1}{\Delta_q}(P_{k1} - P_{k2}) \tag{6.7}$$

with

$$\vartheta_k = \frac{\ln(\mu_{k1}\mu_{k2})}{\Delta_1}, \forall k \in M, l = 1, 2, q = 1, 2, \tag{6.8}$$

and $\tilde{\mu} = max\left\{\frac{1}{\mu_{k1}}, k \in M\right\}$, then, the H_∞ control of system (6.3) is solved under the switching signal

$$\sigma(t) \in \mathcal{T}_{[\Delta_1, \Delta_2]}. \tag{6.9}$$

Proof Let $t_0 = 0$. Denote switching times set $\ell = \{t_0, t_1, t_2, \ldots, t_j, \ldots\}$ generated by $\sigma(t) \in \mathcal{T}_{[\Delta_1,\Delta_2]}$.

When $\sigma(t) = k$, $\forall k \in M$, $t \in [t_j, t_{j+1})$, define

$$\rho(t) = \frac{t - t_j}{t_{j+1} - t_j}, \quad \rho_1(t) = \frac{1}{t_{j+1} - t_j}, \tag{6.10}$$

$$\tilde{\rho}(t) = 1 - \rho(t), \tag{6.11}$$

$$\varphi(t) = (\mu_{k1}\mu_{k2})^{\rho(t)-1}. \tag{6.12}$$

Then, one has $\rho(t) \in [0, 1]$, $\rho(t_j) = 0$, $\rho(t_{j+1}^-) = 1$, and

$$(\mu_{k1}\mu_{k2})^{-1} \le \varphi(t) \le 1, \dot{\varphi}(t) \le \varphi(t)\vartheta_k. \tag{6.13}$$

Choose the time-varying Lyapunov function candidate for subsystem k:

$$V_k(t) = \varphi(t)x^T[\rho(t)P_{k1} + \tilde{\rho}(t)P_{k2}]x \triangleq \varphi(t)x^T P_k(t)x, \tag{6.14}$$

which satisfies

$$\frac{\lambda_2}{\mu}\|x\|^2 \le V_k(t) \le \lambda_1\|x\|^2 \tag{6.15}$$

with $\mu = \max\{\mu_{k1}\mu_{k2}, k \in M\}$, $\lambda_1 = \max\{\lambda_{\max}(P_{kl}), k \in M, l = 1, 2\}$, and $\lambda_2 = \min\{\lambda_{\min}(P_{kl}), k \in M, l = 1, 2\}$.

Now, define a function $\rho_2(t) \in [0, 1]$, such that

$$\rho_1(t) = \frac{1}{\Delta_1}\tilde{\rho}_2(t) + \frac{1}{\Delta_2}\rho_2(t), \tag{6.16}$$

where $\rho_1(t)$ is defined in (6.10). Based on $\rho_2(t)$, let

$$\tilde{\rho}_2(t) = 1 - \rho_2(t). \tag{6.17}$$

In fact, when $\Delta_2 > \Delta_1$, we can simply choose

$$\rho_2(t) = \left(\frac{1}{\Delta_1} - \frac{1}{t_{j+1} - t_j}\right) / \left(\frac{1}{\Delta_1} - \frac{1}{\Delta_2}\right). \tag{6.18}$$

Let $X = [x^T, w^T]^T$. When $t \in [t_j, t_{j+1})$, differentiating $V_k(t)$ along the trajectory of (6.3) yields

$$
\begin{aligned}
&\dot{V}_k(t)\\
&= x^T\left[\dot{\varphi}(t)P_k(t) + \varphi(t)A_k^T P_k(t) + \varphi(t)P_k(t)A_k\right]x\\
&\quad + x^T\left[\varphi(t)\rho_1(t)\left(P_{k1} - P_{k2}\right)\right]x + 2\varphi(t)x^T P_k(t)E_k w\\
&\leq \varphi(t)x^T\left[\vartheta_k P_k(t) + A_k^T P_k(t) + P_k(t)A_k\right]x + \varphi(t)x^T\left[\rho_1(t)\left(P_{k1} - P_{k2}\right)\right]x\\
&\quad + 2\varphi(t)x^T P_k(t)E_k w\\
&= \varphi(t)x^T\left[\vartheta_k\left(\rho(t)P_{k1} + \tilde{\rho}(t)P_{k2}\right)\right]x + \varphi(t)x^T\left[A_k^T\left(\rho(t)P_{k1} + \tilde{\rho}(t)P_{k2}\right)\right]x\\
&\quad + \varphi(t)x^T\left(\rho(t)P_{k1} + \tilde{\rho}(t)P_{k2}\right)A_k x + \varphi(t)x^T\left[\rho_1(t)\left(P_{k1} - P_{k2}\right)\right]x\\
&\quad + 2\varphi(t)x^T\left(\rho(t)P_{k1} + \tilde{\rho}(t)P_{k2}\right)E_k w\\
&\leq \varphi(t)X^T\left\{\rho(t)\left[\rho_2(t)\theta_{k12} + \tilde{\rho}_2(t)\theta_{k11}\right]\right\}X\\
&\quad + \varphi(t)X^T\left\{\tilde{\rho}(t)\left[\rho_2(t)\theta_{k22} + \tilde{\rho}_2(t)\theta_{k21}\right]\right\}X + \gamma^2\varphi(t)w^T w\\
&\leq \lambda\varphi(t)x^T\left(\rho(t)P_{k1} + \tilde{\rho}(t)P_{k2}\right)x + \gamma^2\varphi(t)w^T w\\
&\quad - \varphi(t)\mu_{k1}\mu_{k2}X^T\begin{bmatrix} C_k^T \\ D_k^T \end{bmatrix}\left[\, C_k \;\; D_k \,\right]X\\
&= \lambda\varphi(t)x^T\left(\rho(t)P_{k1} + \tilde{\rho}(t)P_{k2}\right)x - \varphi(t)\mu_{k1}\mu_{k2}y^T y + \gamma^2\varphi(t)w^T w\\
&= \lambda V_k(t) - \varphi(t)\mu_{k1}\mu_{k2}y^T y + \gamma^2\varphi(t)w^T w.
\end{aligned}
\tag{6.19}
$$

By (6.13), one has

$$
-\mu_{k1}\mu_{k2} \leq -\varphi(t)\mu_{k1}\mu_{k2} \leq -1,
$$

which together with (6.19) yields

$$
\dot{V}_k(t) \leq \lambda V_k(t) - y^T y + \gamma^2\varphi(t)w^T w. \tag{6.20}
$$

Note that $\varphi(t_j^-) = 1$ and $\varphi(t_j) = \left(\mu_{\sigma(t_j)1}\mu_{\sigma(t_j)2}\right)^{-1}$ at switching instant t_j. From (6.4), we have

$$
\begin{aligned}
&V_{\sigma(t_j)}(t_j^+)\\
&= \varphi(t_j^+)x^T(t_j^+)\left[\rho(t_j^+)P_{\sigma(t_j)1} + \tilde{\rho}(t_j^+)P_{\sigma(t_j)2}\right]x(t_j^+)\\
&= \left(\mu_{\sigma(t_j)1}\mu_{\sigma(t_j)2}\right)^{-1}x^T(t_j^-)P_{\sigma(t_j)2}x(t_j^-)\\
&\leq \frac{1}{\mu_{\sigma(t_j)1}}x^T(t_j^-)P_{\sigma(t_{j-1})1}x(t_j^-)\\
&= \frac{\varphi(t_j^-)}{\mu_{\sigma(t_j)1}}x^T(t_j^-)\left[\rho(t_j^-)P_{\sigma(t_{j-1})1} + \tilde{\rho}(t_j^-)P_{\sigma(t_{j-1})2}\right]x(t_j^-)\\
&\leq \tilde{\mu}V_{\sigma(t_{j-1})}(t_j^-).
\end{aligned}
\tag{6.21}
$$

It should be noted that $\tilde{\mu} < 1$ in (6.21).

Define $\Gamma(s) = y^T(s)y(s) - \gamma^2 w^T(s)w(s)$ and let $x(t_0) = 0$. When $t \in \left[t_j, t_{j+1}\right)$, by (6.20) and (6.21), we obtain by induction:

$$
\begin{aligned}
& V_{\sigma(t_i)}(t) \\
& \leq V_{\sigma(t_j)}(t_j)e^{\lambda(t-t_j)} - \int_{t_j}^{t} e^{\lambda(t-s)}\Gamma(s)ds \\
& \leq \tilde{\mu}V_{\sigma(t_{j-1})}(t_j^-)e^{\lambda(t-t_j)} - \int_{t_j}^{t} e^{\lambda(t-s)}\Gamma(s)ds \\
& \leq \tilde{\mu}\big[V_{\sigma(t_{j-1})}(t_{j-1})e^{\lambda(t-t_{j-1})} - \int_{t_{j-1}}^{t_j} e^{\lambda(t_j-s)}\Gamma(s)ds\big]e^{\lambda(t-t_j)} \\
& \quad - \int_{t_j}^{t} e^{\lambda(t-s)}\Gamma(s)ds \\
& \leq \cdots \\
& \leq \tilde{\mu}^j e^{\lambda t}V_{\sigma(t_0)}(t_0) - \tilde{\mu}^j \int_{t_0}^{t_1} e^{\lambda(t-s)}\Gamma(s)ds - \tilde{\mu}^{j-1}\int_{t_1}^{t_2} e^{\lambda(t-s)}\Gamma(s)ds - \cdots \\
& \quad -\tilde{\mu}^0 \int_{t_j}^{t} e^{\lambda(t-s)}\Gamma(s)ds \\
& = \tilde{\mu}^{N_\sigma(0,t)}e^{\lambda t}V_{\sigma(t_0)}(t_0) - \int_0^t \tilde{\mu}^{N_\sigma(s,t)}e^{\lambda(t-s)}\Gamma(s)ds \\
& = -\int_0^t e^{\lambda(t-s)+N_\sigma(s,t)\ln\tilde{\mu}}\Gamma(s)ds.
\end{aligned}
\tag{6.22}
$$

Since $V_{\sigma(t_i)}(t) \geq 0$, then, from (6.22), we obtain that

$$
\int_0^t e^{\lambda(t-s)+N_\sigma(s,t)\ln\tilde{\mu}}\Gamma(s)ds \leq 0. \tag{6.23}
$$

Thus, integrating both sides of (6.23) from $t = 0$ to ∞ yields

$$
\int_0^\infty \left(\int_0^t e^{\lambda(t-s)+N_\sigma(s,t)\ln\tilde{\mu}}\Gamma(s)ds\right)dt \leq 0. \tag{6.24}
$$

It should be noted that the double-integral area of (6.24) is

$$
\{(s,t) : 0 \leq t \leq \infty, 0 \leq s \leq t\}.
$$

Now, we rearrange the double-integral area of (6.24), that is,

$$
\{(s,t) : 0 \leq s \leq \infty, s \leq t \leq \infty\},
$$

and obtain

$$
\begin{aligned}
0 &\geq \int_0^\infty \left(\int_0^t e^{\lambda(t-s)+N_\sigma(s,t)\ln\tilde{\mu}} \Gamma(s)ds \right) dt \\
&= \int_0^\infty \Gamma(s)ds \int_s^\infty e^{\lambda(t-s)+N_\sigma(s,t)\ln\tilde{\mu}} dt.
\end{aligned} \tag{6.25}
$$

Define $F(s) = \int_s^\infty e^{\lambda(t-s)+N_\sigma(s,t)\ln\tilde{\mu}} dt$. By (6.6) and $\frac{t-s}{\Delta_2} \leq N_\sigma(s,t) \leq \frac{t-s}{\Delta_1}$, we have

$$
F(s) \leq \int_s^\infty e^{\left(\lambda+\frac{\ln\tilde{\mu}}{\Delta_2}\right)(t-s)} dt = \frac{-1}{\lambda + \frac{\ln\tilde{\mu}}{\Delta_2}}, \tag{6.26}
$$

$$
F(s) \geq \int_s^\infty e^{\left(\lambda+\frac{\ln\tilde{\mu}}{\Delta_1}\right)(t-s)} dt = \frac{-1}{\lambda + \frac{\ln\tilde{\mu}}{\Delta_1}} > 0. \tag{6.27}
$$

That is,

$$
0 < \frac{-1}{\lambda + \frac{\ln\tilde{\mu}}{\Delta_1}} \leq F(s) \leq \frac{-1}{\lambda + \frac{\ln\tilde{\mu}}{\Delta_2}}.
$$

Thus, we have

$$
\int_0^\infty \Gamma(s)ds = \int_0^\infty \|y(s)\|^2 - \gamma^2 \|w(s)\|^2 ds \leq 0. \tag{6.28}
$$

When $w \equiv 0$, by Theorem 1 in [35], switched system (6.3) is GAS under $\sigma(t) \in \mathcal{T}_{[\Delta_1, \Delta_2]}$, since $\tilde{\mu} < 1$. Thus, the proof is completed.

Remark 6.2 In our results, the common $\tilde{\mu}$ and λ are used. Since these two parameters may be dependent on individual subsystems, values of these two parameters can be different for different subsystems, i.e., they can be mode-dependent. This makes our proposed methods less conservative.

To check the conditions of Theorem 6.17, we set

$$
P_{kl} > I, k \in M, l = 1, 2. \tag{6.29}
$$

Hereby, (6.5) is rewritten as

$$
\Theta_{klq}^* = \begin{bmatrix} \theta_{klq}^{(1,1)} - \lambda I & P_{kl}E_k & C_k^T \\ E_k^T P_{kl} & -\gamma^2 I & D_k^T \\ C_k & D_k & -\frac{1}{\mu_{k1}\mu_{k2}} I \end{bmatrix} < 0. \tag{6.30}
$$

With fixed constants $\mu_{kl} > 1, k \in M, l = 1, 2$, and a given specific maximal dwell time Δ_2, the lower bound for admissible minimal dwell time Δ_1 can be defined as

$$
\Delta_1^* = \min_{\Delta_1 \leq \Delta_2} \{\Delta_1 : (6.4), (6.6), (6.29), \text{ and } (6.30) \text{ hold}\}. \tag{6.31}
$$

Then, the H_∞ control for switched system (6.3) is solved under $\sigma(t) \in \mathcal{T}_{[\Delta_1^*, \Delta_2]}$.

Motivated by [35], the following Algorithm 6.1 is designed to compute the admissible minimal dwell time Δ_1^* and its corresponding maximal dwell time Δ_2 with a proper μ_{kl}.

Algorithm 6.1 Computation on μ_{kl}, Δ_1^* and Δ_2

Initialize $\mu_{kl} = \mu_0 > 1$, $\Delta_2 = \Delta_2^* > 0$, $\Delta\mu_1 > 0$, $\epsilon = 10^{-3}$ and $r > 1$;
while $\Delta_2 > \epsilon$ **do**
  Set $N = 0$;
  while $\mu_{kl} > 1$ **do**
    Set $N = N + 1$ and $\mu_{kl} = \mu_{kl} - \Delta\mu_1$;
    Solve (6.31) to obtain Δ_1^*;
    if (6.31) is feasible **then**
      Record $D(N) = \Delta_1^*$;
    else
      Record $D(N) = \Delta_2 + 0.1$;
    end if
  end while
  if $\mu_{kl} > 1$ and $\exists n = 1, 2, \ldots, N$ such that $D(n) \leq \Delta_2$ **then**
    $\Delta_1^* = \min_{n=1,2,\ldots,N}\{D(n)\}$; **Break;**
  else
    Set $\Delta_2 = \Delta_2 / r$;
  end if
  $\mu_{kl} = \mu_0$
end while

Remark 6.3 In order to achieve the internal stabilization, we utilize the time-varying Lyapunov function $V_k(t)$ in (6.14), which was introduced for the first time in [37, 38] for studying the impulsive systems and can also be viewed as a particular case of that considered in [8]. In fact, $V_k(t)$ in (6.14) can be constructed by using multiple matrices $P_{k1}, \ldots, P_{kN}$, as done in [9, 20, 34, 35], which is more general and thus less conservative. However, in order to simplify the mathematical derivations and clearly show the essence of our proposed method, two matrices P_{k1} and P_{k2} are selected to construct the time-varying Lyapunov function. In addition, in our time-varying Lyapunov function $V_k(t)$, there is a time-varying function $\varphi(t)$, which also can introduce a certain degree of freedom.

Remark 6.4 When restricted to the stability analysis case, the main difference of our paper compared to [8] is that *at every time point* within the range of dwell time the LMI conditions in [8] have to be satisfied (see Eq. 54 in Theorem 13 or Eqs. 57–59 in Theorem 15 of [8]) while our proposed conditions need to be satisfied *at only interval ends* of the range of the dwell time.

Remark 6.5 When $\mu_{k1}\mu_{k2} = 1$, $V(t)$ is reduced to the case considered in [35, 39]. The parameters μ_{k1} and μ_{k2} introduce a certain degree of freedom that can be exploited to relax conservativeness (see Sect. 6.6.1).

Due to the fact that switching among stable systems can result in instability, it is necessary to look into the case where all subsystems of switched systems are internally stable. In particular, if the subsystems of switched linear system (6.1) are internally stable, then the condition on the upper bound of the dwell-time (i.e, Δ_2 in Theorem 6.17) can be removed. Thus, we have the following corollary.

Corollary 6.6 *For given constants* $\gamma > 0$, $\lambda > 0$, $\Delta_2 \geq \Delta_1 > 0$, $\mu_{k1} \geq 1$, *and* $\mu_{k2} > 1$, $k \in M$, *if there exist matrices* $P_{k1} > 0$ *and* $P_{k2} > 0$, $k \in M$, *such that (6.4) and*

$$\begin{bmatrix} \theta_{kl}^{(1,1)} + \lambda P_{k,l} & P_{k,l}E_k & C_k^T \\ E_k^T P_{k,l} & -\gamma^2 I & D_k^T \\ C_k & D_k & -\frac{1}{\mu_{k,1}\mu_{k,2}} I \end{bmatrix} < 0, \tag{6.32}$$

$$P_{k1} \geq P_{k2},\ k \in M, \tag{6.33}$$

where

$$\theta_{klq}^{(1,1)} = \vartheta_k P_{k,l} + A_k^T P_{k,l} + P_{k,l} A_k + \frac{1}{\tau_1}\left(P_{k,1} - P_{k,2}\right). \tag{6.34}$$

Then, the H_∞ *control of switched system (6.3) is solved under* $\sigma(t) \in \mathcal{T}_{[\Delta_1,\infty)}$.

Proof Consider the following time-varying Lyapunov function $V_k(t)$ in (6.14). When $t \in \left[t_j, t_{j+1}\right)$, differentiating $V_k(t)$ along the trajectory of (6.3) gives

$$\begin{aligned} &\dot{V}_k(t) \\ &\leq \varphi(t)x^T\left[\vartheta_k P_k(t) + A_k^T P_k(t) + P_k(t)A_k\right]x \\ &\quad + \varphi(t)\rho_1(t)x^T\left(P_{k,1} - P_{k,2}\right)x + 2\varphi(t)x^T P_k(t)E_k w \\ &\leq \varphi(t)\rho(t)x^T\left[\vartheta_k P_{k,1} + A_k^T P_{k,1} + P_{k,1}A_k\right]x \\ &\quad + \varphi(t)\rho(t)x^T\left[\frac{P_{k,1} - P_{k,2}}{\Delta_1} + \gamma^{-2}P_{k,1}E_kE_k^T P_{k,1}\right]x \\ &\quad + \varphi(t)\tilde{\rho}(t)x^T\left[\vartheta_k P_{k,2} + A_k^T P_{k,2} + P_{k,2}A_k\right]x \\ &\quad + \varphi(t)\tilde{\rho}(t)x^T\left[\frac{P_{k,1} - P_{k,2}}{\Delta_1} + \gamma^{-2}P_{k,2}E_kE_k^T P_{k,2}\right]x \\ &\quad + \gamma^2\varphi(t)w^T w \\ &\leq -\lambda V_k(t) - \varphi(t)\mu_{k,1}\mu_{k,2}y^T y + \gamma^2\varphi(t)w^T w, \end{aligned} \tag{6.35}$$

where the second inequality follows from (6.33) and the fact that $2x^T P_{k,l}E_k w \leq \gamma^{-2}x^T P_{k,l}E_kE_k^T P_{k,l}x + \gamma^2 w^T w$, and the third inequality follows from (6.32).

Stability analysis with $w \equiv 0$: From (6.4) and (6.35), we have

$$\dot{V}_k(t) \leq -\lambda V_k(t),\ V_{\sigma(t_j)}(t_j^+) \leq \tilde{\mu} V_{\sigma(t_{j-1})}(t_j^-),$$

where $\tilde{\mu} = max\left\{\frac{1}{\mu_{k1}}, k \in M\right\} < 1$, which means that the Lyapunov function values of the whole switched system (6.3) strictly decreases along the trajectories of switched system (6.3). Then, switched system (6.3) is GAS under $\sigma(t) \in \mathcal{T}_{[\Delta_1,\infty)}$.

L_2-gain analysis: It is immediate from the proof of Theorem 6.17 and thus omitted.

In summary, the H_∞ control of switched system (6.3) is solved under $\sigma(t) \in \mathcal{T}_{[\Delta_1,\infty)}$. This completes the proof.

Remark 6.7 When $\mu_{kl} = 1$, $k \in M$, $l = 1, 2$, in Corollary 6.6, the time-varying Lyapunov function becomes a common time-invariant quadratic Lyapunov function for switched system (6.3). Thus, the proposed H_∞ control of switched system (6.3) can be reduced to that of switched system under arbitrary switchings in the literature such as Theorem 1 in [28].

Remark 6.8 When restricted to the stability analysis with all stable subsystems, the range dwell-time techniques in our paper and [8] are reduced to the minimum dwell-time ones. In this case, our result (i.e., Corollary 6.6) is technically same as that of [8, 39] or [40]. Our conditions are more conservative since the "energy" of the overall switched systems is strictly decreasing while that of [8, 39] or [40] is strictly decreasing at only successive switching times. Our conservativeness is caused by additionally considering the standard H_∞ control objective.

6.4 State Feedback H_∞ Control

In this section, we will now consider the forced switched systems (6.1) for which, in the framework of dwell time technique, state feedback controllers are designed for their own spectacular features. Specifically, we will co-design state-feedback controllers for subsystems and a dwell time switching law to solve the H_∞ control problem for the switched systems (6.1).

If at least one of the subsystems of forced switched system (6.1) is solvable for the H_∞ control problem by designing controllers for subsystems, this problem is trivial. Therefore, the H_∞ control problem for the forced individual subsystems is assumed to be unsolvable. In addition, the H_∞ control problem is not solvable by designing switching signal in the framework of dwell time for switched system (6.1) without control inputs. Now, we will prove that under an appropriate condition, the H_∞ control of switched system (6.1) can be achieved by co-designing state-feedback controllers for the subsystems and a dwell time switching law.

Theorem 6.9 *For given positive constants γ, λ, Δ_1 and Δ_2 with $\Delta_2 \geq \Delta_1$, α_{kl} and $\mu_{kl} > 1$, $k \in M$, $l = 1, 2$, if there exist matrices K_k, $P_{k1} > 0$ and $P_{k2} > 0$, $k \in M$, such that*

$$-\mu_{k_2 2}P_{k_1 1}+P_{k_2 2}<0, \forall k_1,k_2\in M,\ k_1\neq k_2, \tag{6.36}$$

$$\begin{bmatrix} \varpi_{klq}^{(1,1)} & P_{kl}E_k & C_k^T-K_k^TF_k^T & \varpi_{klq}^{(1,3)} \\ * & -\gamma^2 I & D_k^T & 0 \\ * & * & \frac{-1}{\mu_{k1}\mu_{k2}}I & 0 \\ * & * & * & -2\alpha_{kl}I \end{bmatrix}<0, \tag{6.37}$$

$$ln\tilde{\mu}+\lambda\Delta_2<0, \tag{6.38}$$

where

$$\varpi_{klq}^{(1,1)}=\vartheta_k P_{kl}+A_k^TP_{kl}+P_{kl}A_k+\frac{1}{\Delta_q}(P_{k1}-P_{k2})$$
$$-B_kK_k-K_k^TB_k^T-\lambda P_{kl}, \tag{6.39}$$
$$\varpi_{klq}^{(1,3)}=P_{k,l}-I-\alpha_{k,l}(B_kK_k)^T \tag{6.40}$$

with $\vartheta_k=\frac{\ln(\mu_{k1}\mu_{k2})}{\Delta_1}$, $\forall k\in M$, $l=1,2$, $q=1,2$, *and* $\tilde{\mu}=max\left\{\frac{1}{\mu_{k1}},k\in M\right\}$. *Then, the* H_∞ *control of switched system (6.1) is solvable under* $\sigma(t)\in\mathcal{T}_{[\Delta_1,\Delta_2]}$ *and the controllers for individual subsystem k are*

$$u_k=-K_kx. \tag{6.41}$$

Proof We will now prove that the closed-loop switched system satisfies the conditions of Theorem 6.17. Given (6.36) and (6.38) are exactly same as (6.4) and (6.6), respectively, only (6.5) in Theorem 6.17 needs to be satisfied. To achieve this, we first pre- and post-multiply both sides of (6.37) by

$$\begin{bmatrix} I & 0 & 0 & -K_k^TB_k^T \\ 0 & I & 0 & 0 \\ 0 & 0 & I & 0 \end{bmatrix} \tag{6.42}$$

and its transpose, and then obtain that

$$\begin{bmatrix} \upsilon_{klq} & P_{kl}E_k & C_k^T \\ E_k^TP_{kl} & -\gamma^2 I & D_k^T \\ C_k & D_k & -\frac{1}{\mu_{k1}\mu_{k2}}I \end{bmatrix}<0, \tag{6.43}$$

with

$$\upsilon_{klq}=\vartheta_kP_{kl}+(A_k-B_kK_k)^TP_{kl}+P_{kl}(A_k-B_kK_k)$$
$$+\frac{1}{\Delta_q}(P_{k1}-P_{k2})-\lambda P_{kl}. \tag{6.44}$$

which is consistent with inequality (6.5) in which A_k and C_k are replaced by $A_k - B_k K_k$ and $C_k - F_k K_k$, respectively. Therefore, by Theorem 6.17, the H_∞ control problem is solvable by controller (6.41) and the switching signal $\sigma(t) \in \mathcal{T}_{[\Delta_1, \Delta_2]}$.

Remark 6.10 For the stabilization case, besides the proposed method has the similar feature, as it is for the stability analysis (i.e., the proposed LMI conditions in our paper need to be satisfied *at only interval ends* of the range of dwell time), the controller gains in [8] are time-dependent while the proposed controller gains in our paper are time-independent.

Remark 6.11 The stabilization results in [8, 40] are equivalent to the analysis conditions, while ours is sufficient, which is because we pre- and post-multiply both sides of (6.37) by matrix (6.42) and its transpose. Compared to the stabilization results in [8, 40], our result is more conservative. Our controller gains, however, are time-independent, which can be a potential advantage in terms of implementation.

Remark 6.12 Theorem 6.9 can be easily extended to deal with standard H_∞ control of switched system (6.1) by dynamic state feedback control and switched-observers-based output feedback control.

6.5 H_∞ Control for Unforced Switched Linear Systems with Time-Varying Delay

Consider the switched linear systems with time-varying delay:

$$\begin{aligned} \dot{x} &= A_{\sigma(t)}x + A_{d\sigma(t)}x\left(t - \tau(t)\right) + B_{\sigma(t)}w, \\ x\left(\theta\right) &= \phi\left(\theta\right), \theta \in [-r, 0], \\ y &= C_{\sigma(t)}x + D_{\sigma(t)}w, \end{aligned} \tag{6.45}$$

where $x \in R^n$ is state, $y(t) \in R^m$ is controlled output, w is disturbance input which belongs to $L_2[0, +\infty)$, $\sigma(t) : [0, \infty) \to M = \{1, 2, \ldots, m\}$ is the switching signal, A_i, A_{di}, B_i, C_i, D_i, $\forall i \in M$, are constant matrices of appropriate dimension; $\phi(\theta)$ is a differentiable vector-valued initial function on $[-r, 0]$, $r > 0$; $\tau(t)$ denotes the time-varying delay satisfying:

$$0 \le \tau(t) \le r, \dot{\tau}(t) \le d < 1, \tag{6.46}$$

for a known constant d.

The H_∞ performance problem [26, 28, 41] for the switched system (6.45) is stated as follows:

Lemma 6.13 *[42] For given positive scalars $\mu \ge 1$, a and b, which satisfy $0 < b < \frac{a\mu}{\mu+1}$, define*

$$v = \theta(a, b, \mu) \triangleq \frac{1}{\sqrt{\frac{a^2}{4} - \frac{b^2}{\mu}}} \operatorname{arctan} h \frac{(\mu - 1)\sqrt{\frac{a^2}{4} - \frac{b^2}{\mu}}}{\frac{a}{2}(\mu + 1) - 2b}.$$

Let $\varphi(t)$ be the solution of the following initial value problem

$$\dot{\varphi} = -\frac{v}{T}\left(\varphi^2 - a\varphi + \frac{b^2}{\mu}\right), t \geq \alpha, \quad \varphi(\alpha) = \frac{b}{\mu}, \tag{6.47}$$

where $T > 0$. Then $\varphi(t)$ exists on $[\alpha, \infty)$ and satisfies

$$\varphi(\alpha + T) = b, \dot{\varphi}(t) \geq 0, \forall t \geq \alpha. \tag{6.48}$$

Lemma 6.14 *[43]. For any constant matrix $M > 0$, scalars r_1, r_2 satisfying $r_1 < r_2$, and a vector function $\omega : [r_1, r_2] \to R^n$ such that the integrations concerned are well defined, then:*

$$-(r_2 - r_1)\int_{r_1}^{r_2} \omega^T(s) M \omega(s)\, ds \leq -\left(\int_{r_1}^{r_2} \omega^T(s) ds\right)^T M \left(\int_{r_1}^{r_2} \omega(s)\, ds\right). \tag{6.49}$$

6.5.1 Delay-Dependent Dwell Time

Theorem 6.15 *Consider switched system (6.45) with time delay satisfying condition (6.46). For given constants $\gamma > 0$ and $\tau_D > 0$, if there exist scalars $\mu_i \geq 1, a_i, b$ satisfying $0 < b < \frac{a_i \mu_i}{\mu_i + 1}$, $n \times n$ matrices $P_i > 0$, $R_i > 0$, $Q > 0$, $Z > 0$, $n \times 2n$ matrices H_i, $2n \times 2n$ matrices $S_i \geq 0$, $T_i \geq 0$, such that the following hold:*

$$\frac{P_i}{\mu_i} \leq P_j, \tag{6.50}$$

$$\begin{bmatrix} -\frac{v_i}{\tau_D} P_i - R_i & H_i \\ * & \Xi_{i1} \end{bmatrix} \leq 0, \tag{6.51}$$

$$\tilde{\Theta}_i = \begin{bmatrix} \Gamma_{i1} & \Gamma_{i2} \\ * & r B_i^T Z B_i + D_i^T D_i - \gamma^2 I \end{bmatrix} < 0, \tag{6.52}$$

$\forall i, j \in M$ with $i \neq j$, then, switched system (6.45) with any bounded delay $\tau(t)$ satisfying (6.46) has the H_∞ performance with γ under the switching signal $\sigma(t) \in T_{\tau_D}$, where

$$\Xi_{i1} = I_1^T \left[Q - \frac{b^2 v_i}{\tau_D \mu_i} P_i - \frac{b^2}{\mu_i} R_i\right] I_1 - (1 - d) I_2^T Q I_2 + b S_i - \frac{b}{\mu_i} T_i + \Delta_i,$$

$$\Gamma_{i1} = \frac{b}{\mu_i}\Xi_{i2} + \begin{bmatrix} C_i^T C_i & 0 \\ * & 0 \end{bmatrix}, \Gamma_{i2} = \begin{bmatrix} bP_i B_i + rA_i^T Z B_i + C_i^T D_i \\ rA_{di}^T Z B_i \end{bmatrix}$$

with

$$\Delta_i = r\begin{bmatrix} A_i^T \\ A_{di}^T \end{bmatrix} Z \begin{bmatrix} A_i & A_{di} \end{bmatrix} - \frac{1}{r}\begin{bmatrix} I \\ -I \end{bmatrix} Z \begin{bmatrix} I & -I \end{bmatrix}$$

with

$$I_1 = \begin{bmatrix} I & 0 \end{bmatrix}, I_2 = \begin{bmatrix} 0 & I \end{bmatrix}$$

and

$$\begin{aligned}\Xi_{i2} = {} & I_1^T \left[\frac{a_i v_i}{\tau_D} P_i + \frac{b(1+\mu_i)}{\mu_i} R_i\right] I_1 - S_i + T_i \\ & + He\{H_i^T I_1\} + I_1^T He\{P_i A_i\} I_1 + He\{I_1^T P_i A_{di} I_2\}.\end{aligned}$$

Proof For any $i \in M$, $\varphi_i(t)$ denotes the unique solution of the following initial value problem:

$$\begin{cases} \dot{\varphi}_i = -\frac{v_i}{t_{i+1}-t_i}\left(\varphi_i^2 - a_i\varphi_i + \frac{b^2}{\mu_i}\right), t \in \left[t_i, t_{i+1}\right), \\ \varphi_i(t_i) = \frac{b}{\mu_i}, \end{cases} \tag{6.53}$$

where $v_i = \theta(a_i, b, \mu_i)$, and $\{t_i\}$ is the switching time sequence of system (6.45).

By Lemma 6.13, we obtain that $\dot{\varphi}_i(t) \geq 0$ for $t \in \left[t_i, t_{i+1}\right]$, and $\varphi_i\left(t_{i+1}^-\right) = b$. Then, one can verify that:

$$\frac{b}{\mu_i} = \varphi_i(t_i) \leq \varphi_i(t) \leq \varphi_i\left(t_{i+1}^-\right) = b, t \in \left[t_i, t_{i+1}\right]. \tag{6.54}$$

When $t_{i+1} - t_i \geq \tau_D$, we have

$$\dot{\varphi}_i \leq -\frac{v_i}{\tau_D}\left(\varphi_i^2 - a_i\varphi_i + \frac{b^2}{\mu_i}\right), t \in \left[t_i, t_{i+1}\right). \tag{6.55}$$

Based on $\varphi_i(t)$ in (6.53), we define a scalar discontinuous function $\varphi : [0, +\infty) \to [0, +\infty)$ as follows:

$$\begin{aligned} &\varphi(t) = \varphi_i(t), t \in (t_i, t_{i+1}), \\ &\varphi(t_i) \triangleq \varphi\left(t_i^+\right) = \varphi_i(t_i), i = 0, 1, 2, \ldots. \end{aligned} \tag{6.56}$$

One can see that $\varphi(t)$ is a piecewise continuously differentiable function on $[0, +\infty)$.

By (6.54) and (6.55), for $t \in (t_i, t_{i+1}), i = 0, 1, 2, \ldots$, we have

$$\frac{b}{\mu_i} = \varphi(t_i) \le \varphi(t) \le \varphi\left(t_{i+1}^-\right) = b, \tag{6.57}$$

$$\dot{\varphi}(t) \le -\frac{v_i}{\tau_D}\left(\varphi^2(t) - a_i\varphi(t) + \frac{b^2}{\mu_i}\right). \tag{6.58}$$

To study the H_∞ performance of system (6.45), we consider the following piecewise time-varying Lyapunov functional:

$$V(t) = \varphi(t)x^T P_{\sigma(t)}x + \int_{t-\tau(t)}^{t} x^T(s)Qx(s)ds + \int_{-r}^{0}\int_{t+\theta}^{t} \dot{x}^T(s)Z\dot{x}(s)dsd\theta, \tag{6.59}$$

where $P_{\sigma(t)} > 0$, $Q > 0$, $Z > 0$.

We set $V(t) = V(x_t, t)$ for simplicity. Now, we show that $V(t)$ in (6.59) is non-increasing at switching instants t_i.

With the $\varphi(t)$ defined in (6.56), at switching instant $t = t_i$, one can verify that

$$\begin{aligned}
&V(t_i)\\
&= \varphi(t_i)x^T(t_i)P_i x(t_i) + \int_{t_i-\tau(t_i)}^{t_i} x^T(s)Qx(s)ds + \int_{-r}^{0}\int_{t_i+\theta}^{t_i} \dot{x}^T(s)Z\dot{x}(s)dsd\theta\\
&\le bx^T(t_i)P_{k_{i-1}}x(t_i) + \int_{t_i-\tau(t_i)}^{t_i} x^T(s)Qx(s)ds + \int_{-r}^{0}\int_{t_i+\theta}^{t_i} \dot{x}^T(s)Z\dot{x}(s)dsd\theta\\
&= \varphi\left(t_i^-\right)x^T(t_i)P_{k_{i-1}}x(t_i) + \int_{t_i-\tau(t_i)}^{t_i} x^T(s)Qx(s)ds\\
&\quad + \int_{-r}^{0}\int_{t_i+\theta}^{t_i} \dot{x}^T(s)Z\dot{x}(s)dsd\theta\\
&= V\left(t_i^-\right).
\end{aligned} \tag{6.60}$$

In the following, when $w \equiv 0$, we calculate the derivative of $V(t)$ on $[t_i, t_{i+1})$, $i = 0, 1, 2, \ldots$, and together with Lemma 6.14 to achieve the stability of system (6.45) with $w \equiv 0$.

When $t \in [t_i, t_{i+1})$, $i = 0, 1, 2, \ldots$, the derivative of the overall piecewise time-varying Lyapunov functional is

$$\begin{aligned}
&\dot{V}(t)\\
&= \dot{\varphi}(t)x^T P_i x + 2\varphi(t)x^T P_i\dot{x} + x^T Qx - (1 - \dot{\tau}(t))\, x^T(t-\tau(t))Qx(t-\tau(t))
\end{aligned}$$

$$
\begin{aligned}
&+r\dot{x}^T Z\dot{x} - \int_{t-r}^{t} \dot{x}^T(s)Z\dot{x}(s)ds \\
\le & -\frac{v_i}{\tau_D}\left(\varphi^2(t) - a_i\varphi(t) + \frac{b^2}{\mu_i}\right)x^T P_i x + 2\varphi(t)x^T P_i\left[A_i x + A_{di}x(t-\tau(t))\right] \\
&+x^T Qx - (1-d)x^T(t-\tau(t))Qx(t-\tau(t)) \\
&+r\eta^T(t)\begin{bmatrix} A_i^T \\ A_{di}^T \end{bmatrix} Z\begin{bmatrix} A_i & A_{di} \end{bmatrix}\eta(t) - \int_{t-\tau(t)}^{t} \dot{x}^T(s)Z\dot{x}(s)ds \\
\le & -\frac{v_i}{\tau_D}\left(\varphi^2(t) - a_i\varphi(t) + \frac{b^2}{\mu_i}\right)x^T P_i x + 2\varphi(t)x^T P_i\left[A_i x + A_{di}x(t-\tau(t))\right] \\
&+x^T Qx - (1-d)x^T(t-\tau(t))Qx(t-\tau(t)) \\
&+r\eta^T(t)\begin{bmatrix} A_i^T \\ A_{di}^T \end{bmatrix} Z\begin{bmatrix} A_i & A_{di} \end{bmatrix}\eta(t) - \frac{1}{\tau(t)}\int_{t-\tau(t)}^{t} \dot{x}^T(s)dsZ\int_{t-\tau(t)}^{t} \dot{x}(s)ds \\
\le & -\frac{v_i}{\tau_D}\left(\varphi^2(t) - a_i\varphi(t) + \frac{b^2}{\mu_i}\right)x^T P_i x + 2\varphi(t)x^T P_i\left[A_i x + A_{di}x(t-\tau(t))\right] \\
&+x^T Qx - (1-d)x^T(t-\tau(t))Qx(t-\tau(t)) + r\eta^T(t)\begin{bmatrix} A_i^T \\ A_{di}^T \end{bmatrix} Z\begin{bmatrix} A_i & A_{di} \end{bmatrix}\eta(t) \\
&-\frac{1}{r}\left[x - x(t-\tau(t))\right]^T Z\left[x - x(t-\tau(t))\right],
\end{aligned}
\tag{6.61}
$$

where $\eta^T(t) = \begin{bmatrix} x^T & x^T(t-\tau(t)) \end{bmatrix}$.

By (6.57), for any $n \times n$ matrix $R_i \ge 0$, $n \times 2n$ matrix H_i, and $2n \times 2n$ matrices $S_i \ge 0$, $T_i \ge 0$, satisfying (6.51), we have

$$
0 \le \begin{bmatrix} \varphi(t)x \\ \eta(t) \end{bmatrix}^T \begin{bmatrix} \frac{v_i}{\tau_D}P_i + R_i & H_i \\ * & -\Xi_{i1} \end{bmatrix} \begin{bmatrix} \varphi(t)x \\ \eta(t) \end{bmatrix}, \tag{6.62}
$$

and

$$
\begin{aligned}
0 \le & (b-\varphi(t))\left[\varphi(t) - \frac{b}{\mu_i}\right]x^T R_i x + (b-\varphi(t))\,\eta^T(t)S_i\eta(t) \\
&+\left[\varphi(t) - \frac{b}{\mu_i}\right]\eta^T(t)T_i\eta(t).
\end{aligned}
\tag{6.63}
$$

Combining (6.61)–(6.63), we obtain that

$$
\begin{aligned}
&\dot{V}(t)\\
&\leq -\frac{v_i}{\tau_D}\left(\varphi^2(t) - a_i\varphi(t) + \frac{b^2}{\mu_i}\right)x^T P_i x + 2\varphi(t)x^T P_i[A_i x + A_{di}x(t-\tau(t))]\\
&\quad + x^T Qx - (1-d)x^T(t-\tau(t))Qx(t-\tau(t)) + \eta^T(t)\Delta_i\eta(t)\\
&\leq \varphi^2(t)x^T R_i x + \varphi(t)He\{\eta^T(t)H_i^T x\} + \frac{b^2}{\mu_i}x^T R_i x - b\eta^T(t)S_i\eta(t)\\
&\quad + \frac{b}{\mu_i}\eta^T(t)T_i\eta(t) + \frac{v_i a_i}{\tau_D}\varphi(t)x^T P_i x + \varphi(t)x^T He\{P_i A_i\}x\\
&\quad + 2\varphi(t)x^T P_i A_{di}x(t-\tau(t))\\
&\leq (b + \frac{b}{\mu_i})\varphi(t)x^T R_i x - \varphi(t)\eta^T(t)S_i\eta(t) + \varphi(t)\eta^T(t)T_i\eta(t)\\
&\quad + \varphi(t)He\{\eta^T(t)H_i^T x\} + \frac{v_i a_i}{\tau_D}\varphi(t)x^T P_i x + \varphi(t)x^T He\{P_i A_i\}x\\
&\quad + 2\varphi(t)x^T P_i A_{di}x(t-\tau(t))\\
&= (b + \frac{b}{\mu_i})\varphi(t)\eta^T(t)I_1^T R_i I_1\eta(t) - \varphi(t)\eta^T(t)S_i\eta(t) + \varphi(t)\eta^T(t)T_i\eta(t)\\
&\quad + \varphi(t)\eta^T(t)H_i^T I_1\eta(t) + \varphi(t)\eta^T(t)I_1^T H_i\eta(t) + \frac{v_i a_i}{\tau_D}\varphi(t)\eta^T(t)I_1^T P_i I_1\eta(t)\\
&\quad + \varphi(t)\eta^T(t)I_1^T\left(P_i A_i + A_i^T P_i\right)I_1\eta(t) + 2\varphi(t)\eta^T(t)I_1^T P_i A_{di}I_2\eta(t)\\
&= \varphi(t)\eta^T(t)\left[I_1^T\Omega_{1i}I_1 - S_i + T_i + H_i^T I_1 + I_1^T H_i\right]\eta(t)\\
&\quad + \varphi(t)\eta^T(t)\left[+I_1^T He\{P_i A_i\}I_1 + I_1^T P_i A_{di}I_2 + I_2^T A_{di}^T P_i I_1\right]\eta(t)\\
&\leq \varphi(t)\eta^T(t)\Xi_{i2}\eta(t)\\
&< 0,
\end{aligned}
\tag{6.64}
$$

where $\Omega_{1i} = \frac{a_i v_i}{\tau_D}P_i + \frac{b(1+\mu_i)}{\mu_i}R_i$, Then, by (6.64) together with non-increasing of $V(t)$ at switching instants t_i in (6.60), we know that the system (6.45) with $w \equiv 0$ is GAS under the switching signal $\sigma(t) \in T_{\tau_D}$.

When $w \not\equiv 0$, we will show that the system (6.45) has the L_2-gain γ.

Similar to the previous proof, when $t \in [t_i, t_{i+1})$, along the trajectory of (6.45), one has

$$
\begin{aligned}
&\dot{V}(t)\\
&= \dot{\varphi}(t)x^T P_i x + 2\varphi(t)x^T P_i\dot{x} + x^T Qx - (1-\dot{\tau}(t))x^T(t-\tau(t))Qx(t-\tau(t))\\
&\quad + r\dot{x}^T Z\dot{x} - \int_{t-r}^{t}\dot{x}^T(s)Z\dot{x}(s)ds\\
&\leq -\frac{v_i}{\tau_D}\left(\varphi^2(t) - a_i\varphi(t) + \frac{b^2}{\mu_i}\right)x^T P_i x - (1-d)x^T(t-\tau(t))Qx(t-\tau(t))\\
&\quad + x^T Qx + 2\varphi(t)x^T P_i[A_i x + A_{di}x(t-\tau(t)) + B_i w]
\end{aligned}
$$

$$
\begin{aligned}
&+r\left[A_ix+A_{di}x(t-\tau(t))+B_iw\right]^T Z\left[A_ix+A_{di}x(t-\tau(t))+B_iw\right]\\
&-\int\limits_{t-\tau(t)}^{t}\dot{x}^T(s)Z\dot{x}(s)ds\\
\le\ &\varphi(t)\eta^T(t)\Xi_{i2}\eta(t)+\eta^T(t)\Xi_{i3}\eta(t)+2\varphi(t)x^TP_iB_iw\\
&+r\left[A_ix+A_{di}x(t-\tau(t))\right]^T ZB_iw\\
&+rw^TB_i^TZ\left[A_ix+A_{di}x(t-\tau(t))\right]+rw^TB_i^TZB_iw.
\end{aligned}
\tag{6.65}
$$

Define $\xi^T(t)=[x^T, x^T(t-\tau(t)), w^T]$. From (6.65), we have

$$
\begin{aligned}
&\dot{V}(t)+z^T(t)z(t)-\gamma^2w^Tw\\
\le\ &\varphi(t)\eta^T(t)\Xi_{i2}\eta(t)+2\varphi(t)x^TP_iB_iw\\
&+r\left[A_ix+A_{di}x(t-\tau(t))\right]^T ZB_iw+rw^TB_i^TZ\left[A_ix+A_{di}x(t-\tau(t))\right]\\
&+rw^TB_i^TZB_iw+\left[C_ix+D_iw\right]^T\left[C_ix+D_iw\right]-\gamma^2w^Tw\\
=\ &\varphi(t)\eta^T(t)\Xi_{i2}\eta(t)+x^T\left[bP_iB_i+rA_i^TZB_i+C_i^TD_i\right]w\\
&+w^T\left[bB_i^TP_i+rB_i^TZA_i+D_i^TC_i\right]x+x^T(t-\tau(t))A_{di}^TZB_iw\\
&+rw^TB_i^TZA_{di}x(t-\tau(t))+x^TC_i^TC_ix+w^T\left[rB_i^TZB_i+D_i^TD_i-\gamma^2\right]w\\
=\ &\varphi(t)\eta^T(t)\Xi_{i2}\eta(t)+\xi^T(t)\begin{bmatrix} C_i^TC_i & 0 & bP_iB_i+rA_i^TZB_i+C_i^TD_i\\ * & 0 & rA_{di}^TZB_i\\ * & * & rB_i^TZB_i+D_i^TD_i-\gamma^2I\end{bmatrix}\xi(t)\\
=\ &\varphi(t)\eta^T(t)\Xi_{i2}\eta(t)+\xi^T(t)\Theta_i\xi(t),\\
\le\ &\frac{b}{\mu_i}\eta^T(t)\Xi_{i2}\eta(t)+\xi^T(t)\Theta_i\xi(t)\\
=\ &\xi^T(t)\tilde{\Theta}_i\xi(t),
\end{aligned}
\tag{6.66}
$$

where

$$
\tilde{\Theta}_i=\begin{bmatrix}\tilde{\Theta}_i^{11} & \begin{bmatrix} bP_iB_i+rA_i^TZB_i+C_i^TD_i\\ rA_{di}^TZB_i\end{bmatrix}\\ * & rB_i^TZB_i+D_i^TD_i-\gamma^2I\end{bmatrix}
$$

with

$$
\tilde{\Theta}_i^{11}=\frac{b}{\mu_i}\Xi_{i2}+\begin{bmatrix} C_i^TC_i & 0\\ 0 & 0\end{bmatrix}.
$$

From (6.52) and (6.66), we have that

$$
\dot{V}(t)\le-(y^Ty-\gamma^2w^Tw).
\tag{6.67}
$$

When $t\in\left[t_i,t_{i+1}\right)$, by (6.67), we obtain that

$$\begin{aligned} V(t) &\le V_{\sigma(t_i)}(t_i) - \int_{t_i}^{t} \Gamma(s)ds \\ &\le V\left(t_i^-\right) - \int_{t_i}^{t} \Gamma(s)ds \\ &\le V\left(t_{i-1}\right) - \int_{t_{i-1}}^{t} \Gamma(s)ds \\ &\le V\left(t_{i-1}^-\right) - \int_{t_{i-1}}^{t} \Gamma(s)ds \\ &\le \cdots \le V\left(t_0\right) - \int_{t_0}^{t} \Gamma(s)ds, \end{aligned} \tag{6.68}$$

where $\Gamma(s) = y^T(s)y(s) - \gamma^2 w^T(s)w(s)$.

With the initial condition $V_{\sigma(t_0)}(t_0) = 0$ and $V_{\sigma(t_i)}(t) \ge 0$, from (6.68), we can obtain

$$\int_{t_0}^{t} \Gamma(s)ds \le 0. \tag{6.69}$$

When $t \to \infty$, we have

$$\int_{t_0}^{\infty} \Gamma(s)ds \le 0. \tag{6.70}$$

Then, the system (6.45) has the L_2-gain γ under the switching signal $\sigma(t) \in T_{\tau_D}$.

In summary, the system (6.45) has the H_∞ performance with γ under the switching signal $\sigma(t) \in T_{\tau_D}$. Thus, the proof is completed.

Remark 6.16 In order to choose a dwell time switching signal for the switched system based on Theorem 6.15, we first need to choose some parameters in the statement of Theorem 6.15, and then choose them in a "trial and error" design manner but subject to satisfaction of all the assumptions made.

6.5.2 Delay-Independent Dwell Time

In order to study the delay-independent minimum dwell time, we choose $V(x_t, t)$ in (6.59) with $Z = 0$. That is, the following piecewise time-varying Lyapunov functional:

$$V(x_t, t) = \varphi(t)x^T P_{\sigma(t)}x + \int_{t-\tau(t)}^{t} x^T(s)Qx(s)ds, \tag{6.71}$$

which is same to the one in [42].

Theorem 6.17 *Consider switched system (6.45) with time delay satisfying condition (6.46). For given $\tau_D > 0$ and $\forall i, j \in M$, if there exist scalars $\mu_i \geq 1$, a_i, b satisfying $0 < b < \frac{a_i \mu_i}{\mu_i + 1}$, $n \times n$ matrices $P_i > 0$, $R_i > 0$, $Q > 0$, $n \times 2n$ matrices H_i, $2n \times 2n$ matrices $S_i \geq 0$, $T_i \geq 0$, such that the following matrix inequalities hold:*

$$\frac{P_i}{\mu_i} \leq P_j, \tag{6.72}$$

$$\begin{bmatrix} -\frac{v_i}{\tau_D} P_i - R_i & H_i \\ * & \Xi_{i1} \end{bmatrix} \leq 0, \tag{6.73}$$

$$\tilde{\Theta}_i = \begin{bmatrix} \Gamma_{i1} & \Gamma_{i2} \\ * & D_i^T D_i - \gamma^2 I \end{bmatrix} < 0, \tag{6.74}$$

then, switched system (6.45) is globally exponentially stability for any bounded delay $\tau(t)$ satisfying (6.46) with $w \equiv 0$ and has the L_2-gain with γ under $\sigma(t) \in T_{\tau_D}$, where

$$I_1 = \begin{bmatrix} I & 0 \end{bmatrix}, I_2 = \begin{bmatrix} 0 & I \end{bmatrix},$$

$$\Omega_{1i} = \frac{a_i v_i}{\tau_D} P_i + \frac{b(1+\mu_i)}{\mu_i} R_i, \ \Omega_{2i} = -\frac{b^2 v_i}{\tau_D \mu_i} P_i - \frac{b^2}{\mu_i} R_i + Q,$$

$$\Gamma_{i1} = \frac{b}{\mu_i} \Xi_{i2} + \begin{bmatrix} C_i^T C_i & 0 \\ * & 0 \end{bmatrix}, \ \Gamma_{i2} = \begin{bmatrix} bP_i B_i + C_i^T D_i \\ 0 \end{bmatrix},$$

$$\Xi_{i1} = I_1^T \Omega_{2i} I_1 - (1-d) I_2^T Q I_2 + bS_i - \frac{b}{\mu_i} T_i,$$

$$\Xi_{i2} = I_1^T \Omega_{1i} I_1 - S_i + T_i + He\{H_i^T I_1 + I_1^T P_i A_{di} I_2\} + I_1^T He\{P_i A_i\} I_1.$$

Proof The proof can be easily derived by the methodology as the ones of Theorem 1 in [42] and Theorem 6.15 by setting $\Delta_i = 0$. Therefore, the proof of Theorem 6.17 is omitted in here.

Remark 6.18 Unlike in [30, 33] where only the weighted H_∞ performance is obtained, Theorem 6.17 presents the sufficient conditions to guarantee the standard H_∞ performance by restricting the decay of the Lyapunov functional of the active subsystem and forcing "energy" of the overall switched system to decrease at switching instants by the proposed Lyapunov functionals in (6.59) and (6.71), respectively.

Remark 6.19 Theorem 6.17 proposes the conditions to guarantee the existence of the delay-independent minimum dwell time in the sense that the switched time delay system with such minimum dwell time has the H_∞ performance irrespective of the sizes of time delay. However, the derived minimum (or average) dwell times in [30, 33] depend upon the upper bounds of time delay, which yields that the results in the above-mentioned literature will be difficult to apply to design dwell time switching rules for the switched systems with unknown delay.

Remark 6.20 It should be noted that our results can be extended to study the H_∞ performance of switched system with some uncertainties, for example, as in [42].

In addition, if $A_{di} = 0$ and $r = 0$, then the switched time-varying delay system (6.45) reduces to the following switched system without time-delay:

$$\begin{aligned} \dot{x} &= A_{\sigma(t)}x + B_{\sigma(t)}w, \\ y &= C_{\sigma(t)}x + D_{\sigma(t)}w, \end{aligned} \tag{6.75}$$

Corollary 6.21 *Consider switched system (6.75). For given $\tau_D > 0$ and $\forall i, j \in M$, if there exist scalars $\mu_i \geq 1$, a_i, b satisfying $0 < b < \frac{a_i\mu_i}{\mu_i+1}$, $n \times n$ matrices $P_i > 0$, $R_i > 0$, H_i, $S_i \geq 0$, $T_i \geq 0$, such that the following matrix inequalities hold:*

$$\frac{P_i}{\mu_i} \leq P_j, \tag{6.76}$$

$$\begin{bmatrix} -\frac{v_i}{\tau_D}P_i - R_i & H_i \\ * & bS_i - \frac{b^2 v_i}{\tau_D \mu_i}P_i - \frac{b^2}{\mu_i}R_i - \frac{b}{\mu_i}T_i \end{bmatrix} \leq 0, \tag{6.77}$$

$$\tilde{\Theta}_i = \begin{bmatrix} \Gamma_{i1} & bP_iB_i + C_i^T D_i \\ * & D_i^T D_i - \gamma^2 I \end{bmatrix} < 0, \tag{6.78}$$

where

$$\Gamma_{i1} = \frac{b}{\mu_i}(\frac{a_i v_i}{\tau_D}P_i + \frac{b(1+\mu_i)}{\mu_i}R_i - S_i + T_i + He\{H_i + P_iA_i\}) + C_i^T C_i,$$

then, switched system (6.75) is globally exponentially stability with $w \equiv 0$ and has the L_2-gain with γ under $\sigma(t) \in T_{\tau_D}$.

6.6 Illustrative Examples

6.6.1 Stabilization of Unforced Switched Linear Systems

We consider the example in [35]:

$$\dot{x} = A_\sigma x, \tag{6.79}$$

with $A_1 = \begin{bmatrix} -1.9 & 0.6 \\ 0.6 & -0.1 \end{bmatrix}$, $A_2 = \begin{bmatrix} 0.1 & -0.9 \\ 0.1 & -1.4 \end{bmatrix}$.

Set $\mu_{11} = \mu_{21} = \mu_{12} = \mu_{22} = 1.01$. Then, the following feasible solution can be obtained

$P_{11} = \begin{bmatrix} 5.7966 & -1.9228 \\ -1.9228 & 3.5987 \end{bmatrix}$, $P_{12} = \begin{bmatrix} 2.2864 & 1.2348 \\ 1.2348 & 6.6429 \end{bmatrix}$,

$P_{21} = \begin{bmatrix} 2.2820 & 1.2174 \\ 1.2174 & 6.5847 \end{bmatrix}$, $P_{22} = \begin{bmatrix} 5.7848 & -1.9196 \\ -1.9196 & 3.6243 \end{bmatrix}$.

For the above system with (6.31), we carried out the simulation. Table 6.1 and Fig. 6.1 are the comparison in terms of computation time, which clearly shows our method needs less computation time when the division parameter L (as defined in [35]) is greater than 1. Figure 6.2 is the stability region achieved by our method, i.e., $\bigcup_{i=0,1,2,3,10} R_i$. Compared to the stability region in [35] (i.e., $\bigcup_{i=1,2,3,10} R_i$), the *additional* region by our method is R_0, which is marked in green (Fig. 6.2).

6.6.2 H_∞ Control of Unforced Switched Linear Systems with All Unstable Subsystems

We consider switched system (6.3) with two subsystems as:

Table 6.1 Calculation time for Example A

Method	Numbers of LMIs	Time (s)
[35]	12 (L=1)	0.037294
	18 (L=2)	0.061200
	24 (L=3)	0.077739
	30 (L=4)	0.098330
Our proposed	14	0.057814

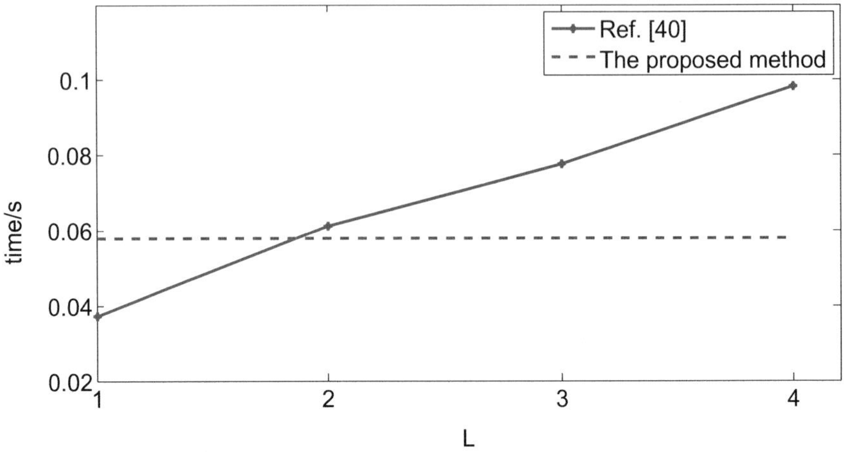

Fig. 6.1 Computational time for Example A

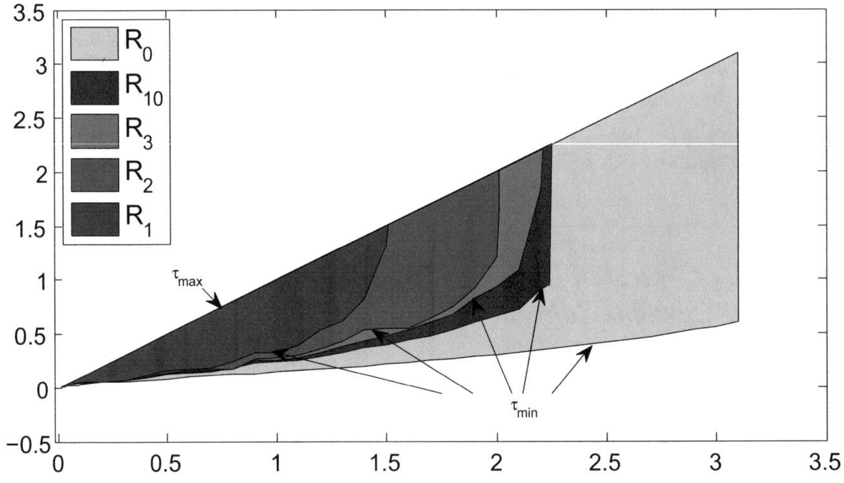

Fig. 6.2 Stability region for admissible dwell time for Example A

$$\left[\begin{array}{c|c} A_1 & E_1 \\ \hline C_1 & D_1 \end{array}\right] = \left[\begin{array}{ccc|c} -2 & -0.1 & 0.1 & 0 \\ -1.2 & -2.4 & -1 & 0.1 \\ -0.1 & 0.01 & 0.02 & -0.1 \\ \hline -0.1 & 0 & 0.1 & 0.5 \end{array}\right],$$

$$\left[\begin{array}{c|c} A_2 & E_2 \\ \hline C_2 & D_2 \end{array}\right] = \left[\begin{array}{ccc|c} -1.5 & -0.4 & -1 & 0 \\ 0.1 & -0.1 & 0.5 & 0.1 \\ 1.5 & 1 & -0.6 & -0.1 \\ \hline 0 & 0.1 & -0.1 & -0.5 \end{array}\right].$$

It should be noted that each subsystem of switched system is *unstable*, since the eigenvalues of A_1 and A_2 are $\{-2.6015, -1.7868, 0.0083\}$ and $\{0.0385, -1.1192\pm 0.9917i\}$, respectively. Therefore, the method in [8] cannot be applied (even for the internal stability) since the necessary condition in [8] is that at least one subsystem be stable, and the methods in [28, 32, 33] are not applicable for the H_∞ control of the studied switched system as well. However, it is still possible to design a set of dwell time switching signals to achieve the H_∞ property by our proposed method for the above switched systems.

Let $\gamma = 0.52$, $\Delta_1 = \Delta_2 = 0.1$, $\lambda = 0.07$, and $\mu_{11} = \mu_{12} = \mu_{21} = \mu_{22} = 1.018$. By Theorem 6.17, the H_∞ performance of switched system (6.3) is achieved under the switching law with the period of 0.1 s.

6.6.3 H_∞ Control of Forced Switched Linear Systems

We consider switched system (6.1) with two subsystems, where

$$A_1 = \begin{bmatrix} -2 & 0 & 0 \\ 1 & 0.4 & 0 \\ 0 & 1 & -0.4 \end{bmatrix}, A_2 = \begin{bmatrix} -1.5 & 1 & 0 \\ 0 & -2 & -1 \\ 0 & -1 & 0.1 \end{bmatrix},$$
$$B_1 = \begin{bmatrix} 0, & 0, & 1 \end{bmatrix}^T, B_2 = \begin{bmatrix} 1, & 0, & 0 \end{bmatrix}^T,$$
$$E_1 = \begin{bmatrix} 0.1, & 0, & -1 \end{bmatrix}^T, E_2 = \begin{bmatrix} 0, & 0.1, & 1 \end{bmatrix}^T,$$
$$C_1 = \begin{bmatrix} 0, & 0.1, & 0.2 \end{bmatrix}, C_2 = \begin{bmatrix} 0.1, & -0.3, & 0 \end{bmatrix},$$
$$D_1 = -D_2 = 0.2, \quad F_1 = F_2 = 0.$$

Based on the controllability decomposition (which can be done by 'ctrbf' in MATLAB), for subsystems 1 and 2, we can obtain that

$$\left(\begin{array}{cc:c} -2 & 0 & 0 \\ 1 & 0.4 & 0 \\ \hdashline 0 & 1 & -0.4 \end{array}\right), \left(\begin{array}{c} 0 \\ 0 \\ \hdashline 1 \end{array}\right), \left(\begin{array}{cc:c} 0.1 & -1 & 0 \\ -1 & -2 & 0 \\ \hdashline 0 & 1 & -1.5 \end{array}\right), \left(\begin{array}{c} 0 \\ 0 \\ \hdashline 1 \end{array}\right).$$

Then the controllability of the studied system is not satisfied, and the eigenvalues of *both* the uncontrollable systems are $\{0.4, -2\}$ and $\{0.5, -2.5\}$, respectively. Therefore, the stabilization of each subsystem cannot be achieved through the state feedback when $w \equiv 0$. Thus, the method in [8] is not applicable to its internal stabilization. Because the proposed method can apply to the case that *none* of subsystems are stabilizable, we can apply it to this example to achieve H_∞ control of the switched system, as shown below.

Let $\gamma = 1.55$, $\lambda = 4.5$, $\Delta_1 = \Delta_2 = 0.03$, $\mu_{11} = \mu_{12} = \mu_{21} = \mu_{22} = 1.15$, and $\alpha_{11} = \alpha_{12} = \alpha_{21} = \alpha_{22} = 1$. By Theorem 6.9, we obtain controller gains for each subsystem

$$K1 = [-0.6247, 0.0106, -2.0216],$$
$$K2 = [-7.0164, 0.5836, -0.5708].$$

It should be noted that each closed-loop subsystem of switched system is unstable, since the eigenvalues of $A_1 - B_1K_1$ and $A_2 - B_2K_2$ are $\{0.4, -2, -2.4216\}$ and $\{0.5, -2.4, -8.5146\}$, respectively. By Theorem 6.9, the H_∞ control is achieved for the switched system under $\sigma(t) \in \mathcal{T}_{[0.03,0.03]}$.

6.6.4 H_∞ Performance of Unforced Switched Linear Systems with Stable Subsystems

We consider the switched system (6.3) with two subsystems, where

$$A_1 = \begin{bmatrix} -1 & 1 & 0 \\ 0 & -1 & 1 \\ 0 & 0 & -4 \end{bmatrix}, \quad A_2 = \begin{bmatrix} -4 & 1 & -1 \\ 0 & -3 & -1 \\ 2 & -2 & -4 \end{bmatrix},$$
$$B_1 = [1,\ 1,\ 0]^T, \quad B_2 = [1,\ 0,\ 1]^T,$$
$$C_1 = [-1,\ 0,\ 1], \quad C_2 = [1,\ 0,\ -1.5], D_1 = -D_2 = 2.$$

Based on our proposed method, when $\Delta_1 = 0.15$ and $\gamma = 2.48$ with additional parameters $\lambda = 0.0001$, and $\mu_{11} = \mu_{12} = \mu_{21} = \mu_{22} = 1.01$, Corollary 6.6 achieves the H_∞ control. It is easy to know that the method in [28] has a larger minimum dwell time, since for $\Delta_1 = 0.15$ and $\gamma = 2.48$ the positive definite stabilizing solution of the algebraic Riccati equation in [28] does not exist.

6.6.5 H_∞ Control for Unforced Switched Linear Systems with Time-Varying Delay

In this subsection, an example is studied to demonstrate the effectiveness of the proposed results for switched system (6.45) composed of two subsystems ($i = 1, 2,$) with the parameters:

$$A_1 = \begin{bmatrix} -1.83 & -2 \\ 0 & -8.91 \end{bmatrix}, A_2 = \begin{bmatrix} -0.71 & 3 \\ 0 & -9.95 \end{bmatrix},$$
$$A_{d1} = \begin{bmatrix} -1 & 1 \\ 0.3 & -4 \end{bmatrix}, A_{d2} = \begin{bmatrix} -2 & 1 \\ 0.3 & -2.7 \end{bmatrix}, B_1 = \begin{bmatrix} -0.2 \\ 0.1 \end{bmatrix}, B_2 = \begin{bmatrix} -0.1 \\ 0.2 \end{bmatrix},$$
$$C_1 = \begin{bmatrix} -0.5 & 0.1 \end{bmatrix}, C_2 = \begin{bmatrix} 0.1 & -1.2 \end{bmatrix}, D_1 = 0.85, D_2 = 0.91.$$

Our purpose is to identify the admissible dwell time switching signals to ensure that the underlying switched system (6.45) is globally asymptotically stable with a prescribed standard H_∞ performance.

Let $\tau_D = 0.1$ and set $r = 0.6, d = 0.7, \gamma = 1.299, \mu_1 = \mu_2 = 1.0002, a_1 = 2.02, a_2 = 2.003$. By Theorem 6.15, we obtain that

$$P_1 = \begin{bmatrix} 6.5008 & 3.5341 \\ 3.5341 & 55.9698 \end{bmatrix}, R_1 = \begin{bmatrix} 112.1086 & 6.0777 \\ 6.0777 & 193.7678 \end{bmatrix},$$

$$P_2 = \begin{bmatrix} 6.5008 & 3.5341 \\ 3.5341 & 55.9698 \end{bmatrix}, R_2 = \begin{bmatrix} 124.2143 & 0.1830 \\ 0.1830 & 216.3390 \end{bmatrix},$$

$$H_1 = \begin{bmatrix} -109.3876 & 4.9690 & -2.6854 & -0.0120 \\ 10.5714 & -52.4857 & -14.4111 & 40.6093 \end{bmatrix},$$

$$H_2 = \begin{bmatrix} -126.4063 & 3.5688 & -0.0452 & -0.3757 \\ 14.8025 & -25.4697 & -5.7702 & 18.4690 \end{bmatrix},$$

$$S_1 = \begin{bmatrix} 158.0438 & -0.0078 & -6.5785 & -4.4702 \\ -0.0078 & 100.4070 & -1.9420 & -24.9564 \\ -6.5785 & -1.9420 & 164.7267 & 3.1782 \\ -4.4702 & -24.9564 & 3.1782 & 196.9081 \end{bmatrix},$$

$$S_2 = \begin{bmatrix} 159.7203 & 2.9160 & -1.8921 & -4.3320 \\ 2.9160 & 99.8777 & -1.1617 & -20.2843 \\ -1.8921 & -1.1617 & 149.5958 & 1.4421 \\ -4.3320 & -20.2843 & 1.4421 & 203.1892 \end{bmatrix},$$

$$T_1 = \begin{bmatrix} 170.8509 & 10.3771 & 4.7455 & 4.9620 \\ 10.3771 & 691.5750 & 4.6838 & 135.9599 \\ 4.7455 & 4.6838 & 160.2930 & 1.8751 \\ 4.9620 & 135.9599 & 1.8751 & 173.3256 \end{bmatrix},$$

$$T_2 = \begin{bmatrix} 165.4265 & -10.7204 & 10.1460 & 0.5831 \\ -10.7204 & 616.0785 & -0.6638 & 90.3061 \\ 10.1460 & -0.6638 & 148.9518 & 1.4849 \\ 0.5831 & 90.3061 & 1.4849 & 155.9150 \end{bmatrix},$$

$$Z = \begin{bmatrix} 5.0769 & 2.3503 \\ 2.3503 & 7.0110 \end{bmatrix}.$$

According to Theorem 6.15, the switched system (6.45) has the H_∞ performance with $\gamma = 1.299$ under the switching signal with dwell time $\tau_D = 0.1$.

When the initial point $x_0 = 0$. Figure 6.3 is the state response of the switched system (6.45), when $w = 0$ under a randomly chosen switching signal with the dwell time $\tau_D = 0.1$ second depicted in Fig. 6.4.

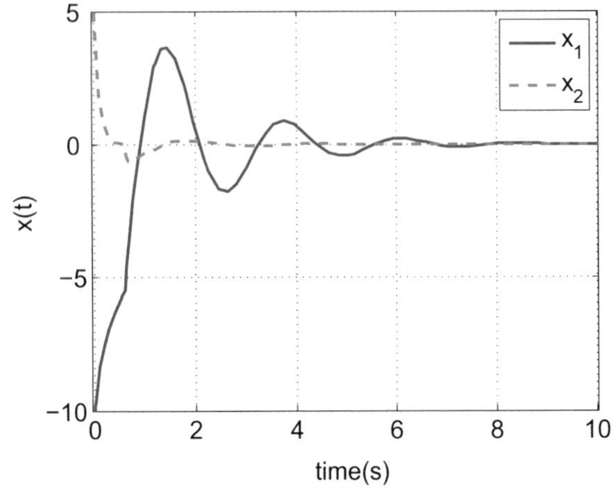

Fig. 6.3 The trajectories of switched system with w

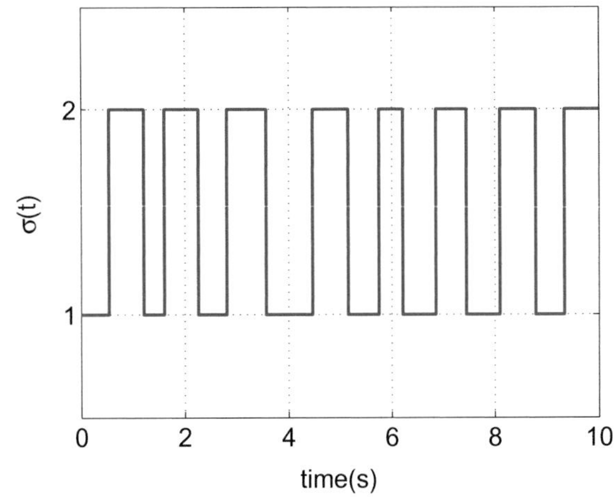

Fig. 6.4 The switching signal

On the other hand, when $w = 0.1(sin(10 + \frac{t}{2}) + 3.1)e^{-0.05t}$ and when the initial point $x_0 = [-10, 5]^T$. Figure 6.5 is the state response of the switched system (6.45), which shows that the switched system (6.45) is asymptotically stable under the switching signal with dwell time $\tau_D = 0.1$ depicted in Fig. 6.6.

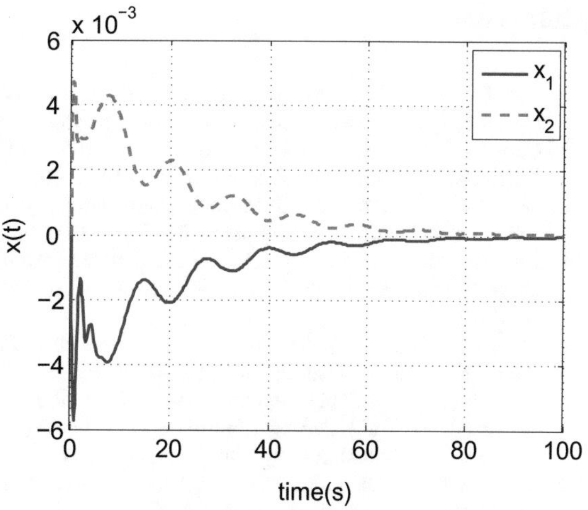

Fig. 6.5 The trajectories of switched system without w

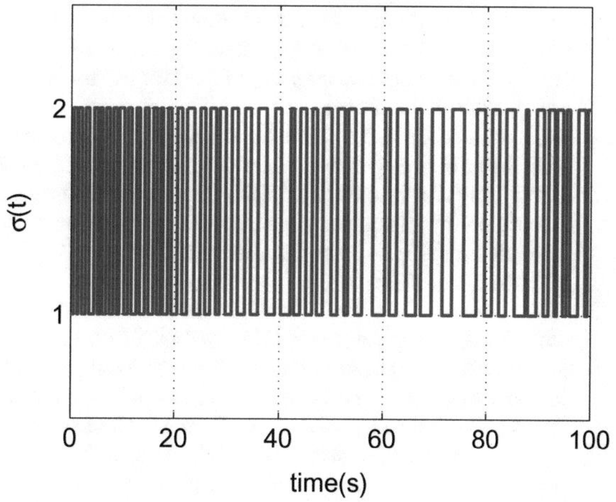

Fig. 6.6 The switching signal

6.7 Concluding Remarks

In this chapter, we have for the first time investigated the standard H_∞ control of switched linear systems via dwell time switchings without posing any internal stability requirements on subsystems. First, for unforced switched linear systems, a type of multiple time-varying Lyapunov functions is utilized, and a sufficient condition is also given to achieve the standard H_∞ control of the unforced switched linear systems. Then, state feedback controllers are designed in the same framework of the dwell time for that of the forced switched linear systems.

References

1. Lan, W., Chen, B.: Explicit construction of h_∞ control law for a class of nonminimum phase nonlinear systems. Automatica **44**(3), 738–744 (2008)
2. Qian, C., Lin, W.: Almost disturbance decoupling for a class of high-order nonlinear systems. IEEE Trans. Autom. Control **45**(6), 1208–1214 (2000)
3. Gumussoy, S., Özbay, H.: Sensitivity minimization by strongly stabilizing controllers for a class of unstable time-delay systems. IEEE Trans. Autom. Control **54**(3), 590 (2009)
4. Khargonekar, P.P., Petersen, I.R., Zhou, K.: Robust stabilization of uncertain linear systems: quadratic stabilizability and h_∞ control theory. IEEE Trans. Autom. Control **35**(3), 356–361 (1990)
5. Chai, T.: An indirect stochastic adaptive scheme with on-line choice of weighting polynomials. IEEE Trans. Autom. Control **35**(1), 82–85 (1990)
6. Branicky, M.S.: Multiple lyapunov functions and other analysis tools for switched and hybrid systems. IEEE Trans. Autom. Control **43**(4), 475–482 (1998)
7. Liberzon, D.: Switching in Systems and Control. Birkhauser, Boston (2003)
8. Briat, C.: Convex conditions for robust stabilization of uncertain switched systems with guaranteed minimum and mode-dependent dwell-time. Syst. Control Lett. **78**, 63–72 (2015)
9. Briat, C.: Stability analysis and stabilization of stochastic linear impulsive, switched and sampled-data systems under dwell-time constraints. Automatica **74**, 279–287 (2016)
10. Colaneri, P., Middleton, R.H., Chen, Z., Caporale, D., Blanchini, F.: Convexity of the cost functional in an optimal control problem for a class of positive switched systems. Automatica **50**(4), 1227–1234 (2014)
11. Fu, J., Ma, R., Chai, T.: Global finite-time stabilization of a class of switched nonlinear systems with the powers of positive odd rational numbers. Automatica **54**(4), 360–373 (2015)
12. Deng, Fang and Qiu, Huangbin and Chen, Jie and Wang, Lu and Wang, Bo: Wearable thermoelectric power generators combined with flexible supercapacitor for low-power human diagnosis devices. IEEE Trans. Ind. Electron. **64**(2), 1477–1485 (2017). https://doi.org/10.1109/TIE.2016.2613063
13. Liu, J., Liu, X., Xie, W.-C.: Input-to-state stability of impulsive and switching hybrid systems with time-delay. Automatica **47**(5), 899–908 (2011)
14. Ma, R., Zhao, J.: Backstepping design for global stabilization of switched nonlinear systems in lower triangular form under arbitrary switchings. Automatica **46**(11), 1819–1823 (2010)
15. Sun, Z., Ge, S.S.: Switched Linear Systems: Control and Design. Springer, London (2005)
16. Yang, H., Jiang, B., Zhao, J.: On finite-time stability of cyclic switched nonlinear systems. IEEE Trans. Autom. Control **60**(8), 2201–2206 (2015)
17. Zhang, L., Zhuang, S., Shi, P.: Non-weighted quasi-time-dependent h_∞ filtering for switched linear systems with persistent dwell-time. Automatica **54**, 201–209 (2015)
18. Yue, D., Han, Q.-L., Lam, J.: Network-based robust h_∞ control of systems with uncertainty. Automatica **41**(6), 999–1007 (2005)
19. Fujita, M., Kawai, H., Spong, M.W.: Passivity-based dynamic visual feedback control for three-dimensional target tracking: stability and L_2-gain performance analysis. IEEE Trans. Control Syst. Technol. **15**(1), 40–52 (2007)
20. Allerhand, L.I., Shaked, U.: Robust state-dependent switching of linear systems with dwell time. IEEE Trans. Autom. Control **58**(4), 994–1001 (2013)
21. Duan, C., Fen, W.: Analysis and control of switched linear systems via dwell-time min-switching. Syst. Control Lett. **70**, 8–16 (2014)
22. Hajiahmadi, M., De Schutter, B., Hellendoorn, H.: Stabilization and robust H_∞ control for sector-bounded switched nonlinear systems. Automatica **50**(10), 2726–2731
23. Long, L., Zhao, J.: H_∞ control of switched nonlinear systems in p-normal form using multiple lyapunov functions. IEEE Trans. Autom. Control **57**(5), 1285–1291 (2012)
24. Sun, X.-M., Liu, G.-P., Wang, W., Rees, D.: L_2-gain of systems with input delays and controller temporary failure: zero-order hold model. IEEE Trans. Control Syst. Technol. **19**(3), 699–706 (2011)

25. Xu, H., Teo, K.L.: Exponential stability with L_2-gain condition of nonlinear impulsive switched systems. IEEE Trans. Autom. Control **55**(10), 2429–2433 (2010)
26. Zhao, J., Hill, D.J.: On stability, L_2-gain and H_∞ control for switched systems. Automatica **44**(5), 1220–1232 (2008)
27. Briat, C.: Convex lifted conditions for robust l_2-stability analysis and l_2-stabilization of linear discrete-time switched systems with minimum dwell-time constraint. Automatica **50**(3), 976–983 (2014)
28. Geromel, J.C., Colaneri, P.: H_∞ and dwell time specifications of continuous-time switched linear systems. IEEE Trans. Autom. Control **55**(1), 207–212 (2010)
29. Hespanha, J.P.: Logic-Based Switching Algorithms in Control. Thesis (1998)
30. Sun, X.-M., Zhao, J., Hill, D.J.: Stability and L_2-gain analysis for switched delay systems: a delay-dependent method. Automatica **42**(10), 1769–1774 (2006)
31. Yuan, C., Wu, F.: Hybrid control for switched linear systems with average dwell time. IEEE Trans. Autom. Control **60**(1), 240–245 (2015)
32. Zhai, G., Hu, B., Yasuda, K., Michel, A.N.: Disturbance attenuation properties of time-controlled switched systems. J. Frankl. Inst. **338**(7), 765–779 (2001)
33. Zhao, X., Liu, H., Wang, Z.: Weighted H_∞ performance analysis of switched linear systems with mode-dependent average dwell time. Int. J. Syst. Sci. **44**(11), 2130–2139 (2013)
34. Briat, C., Seuret, A.: Affine characterizations of minimal and mode-dependent dwell-times for uncertain linear switched systems. IEEE Trans. Autom. Control **58**(5), 1304–1310 (2013)
35. Xiang, W., Xiao, J.: Stabilization of switched continuous-time systems with all modes unstable via dwell time switching. Automatica **50**(3), 940–945 (2014)
36. Ma, R., Fu, J., Chai, T.: Dwell-time-based observers design for unknown inputs switched linear systems without requiring strong detectability of subsystems. IEEE Trans. Autom. Control **62**(8), 4215–4221 (2017)
37. Chen, W.-H., Li, D.-X., Lu, X.: Impulsive functional observers for linear systems. Int. J. Control Autom. Syst. **9**(5), 987–992 (2011)
38. Chen, W.-H., Li, D.-X., Lu, X.: Impulsive observers with variable update intervals for lipschitz nonlinear time-delay systems. Int. J. Syst. Sci. **44**(10), 1934–1947 (2013)
39. Allerhand, I.L., Shaked, U.: Robust stability and stabilization of linear switched systems with dwell time. IEEE Trans. Autom. Control **56**(2), 381–386 (2011)
40. Geromel, J.C., Colaneri, P.: Stability and stabilization of continuous-time switched linear systems. SIAM J. Control Optim. **45**(5), 1915–1930 (2006)
41. Fu, J., Ma, R., Chai, T., Hu, Z.: Dwell-time-based standard h_∞ control of switched systems without requiring internal stability of subsystems. IEEE Trans. Autom. Control 1–1 (2018)
42. Chen, W.-H., Zheng, W.X.: Delay-independent minimum dwell time for exponential stability of uncertain switched delay systems. IEEE Trans. Autom. Control **55**(10), 2406–2413 (2010)
43. Gu, K., Chen, J., Kharitonov, L.V.: Stability of Time-Delay Systems. Birkhäuser, Boston (2003)

Chapter 7
Observers Design for Unknown Inputs Switched Linear Systems Without Requiring Strong Detectability of Subsystems

This chapter investigates state observers design of a class of unknown inputs switched linear systems via mode-dependent dwell time switchings. The distinguishing feature of the proposed method is that strong detectability condition of subsystems of the switched systems is unnecessarily required. Firstly, a time-varying coordinate transformation is introduced to design a suitable reduced-order observer for each subsystem. Then, computable sufficient conditions on the synthesis of the observers are proposed in the framework of a mode-dependent dwell time technique. Since the observer of individual subsystem cannot be designed due to unavailability of strong detectability condition of the subsystem, the state of the switched system is estimated under the condition of confining the dwell time by a pair of upper and lower bounds, restricting the growth of Lyapunov function of the active subsystem, and forcing "energy" of the overall switched system to decrease at *switching instants*. Next, we apply our method to the stabilization of switched singular linear systems. Finally, examples are presented to demonstrate the effectiveness of the proposed methods.

7.1 Introduction

Switched systems have drawn considerable attention because of their importance from both theoretical and practical points of view [1–8]. Main issues on switched systems such as stability [9–13], stabilization [14–16], H_∞ control [17], dissipativity theory [18], and optimal control [19] have been extensively studied. In addition, as full or partial information of state is unavailable for feedback in abundant practical applications [20–22], state estimation is an important issue for such applications. Therefore, this paper will focus on state estimation of switched systems.

For the considered state estimation problem, the most relevant papers are [23–32] which studied state estimation of switched systems under different switching strategies. However, these switched systems do not contain unknown inputs.

J. Fu and R. Ma, *Stabilization and H_∞ Control of Switched Dynamic Systems*, Studies in Systems, Decision and Control 310,
https://doi.org/10.1007/978-3-030-54197-2_7

Since the unknown inputs not only impact the switched systems directly, but also enter the error dynamics from observer design, the methods in [23–28, 33] are not applicable to switched systems containing unknown inputs. To particularly focus on observers design for the unknown inputs switched systems, some methods have been proposed. [29–32] decoupled the unknown inputs from the dynamics of unavailable state component for observer designs *under the assumption that all or part of subsystems are strongly detectable*. This assumption indeed compromises to the obstacle that if none of subsystems is strongly detectable, then it is impossible to design state observers. Thus, questions naturally arise: is it possible, in the framework of the dwell time technique, to achieve the observers design of switched linear systems with unknown inputs *without posing any strong detectability requirements on subsystems of switched systems*? If possible, under what conditions can we achieve this goal and how? To our best knowledge, in the literature there have not been results which provide answers to these questions. This is the motivation of the present paper.

From the motivation above, this paper first applies a time-varying coordinate transformation to each subsystem to design a set of reduced-order observers. Then computable sufficient conditions on the synthesis of the observers are derived in the framework of dwell time techniques. Next, stability of the error dynamics is analyzed under the condition of confining the dwell time by a pair of upper and lower bounds, restricting the growth of Lyapunov function of the active subsystem, and forcing "energy" of the overall switched system to decrease at *switching instants*. Last, we apply our method to the stabilization of switched singular linear systems.

Notations: I and 0 denote an identity matrix and zero matrix of appropriate dimension, respectively. $\Gamma^{\perp}$ is a matrix such that $\Gamma^{\perp}\Gamma=0$ and $\Gamma^{\perp}$ is linearly independent of Γ^T. $\Gamma^{+}=\left[\Gamma^{\perp}\Gamma\right]^{-1}\Gamma^T$ is the left pseudo-inverse of Γ. Denote $x(t_j)=x(t_j^+)=\lim\limits_{h\to 0^+}x(t_j+h)$, $x(t_j^-)=\lim\limits_{h\to 0^-}x(t_j+h)$, i.e., $x(t)$ is right continuous.

7.2 Problem Formulation

We consider the following unknown inputs switched linear systems:

$$\begin{aligned}\dot{x}(t) &= A_{\sigma(t)}x(t)+B_{\sigma(t)}u(t)+E_{\sigma(t)}w(t),\\ y(t) &= C_{\sigma(t)}x(t),\end{aligned}\tag{7.1}$$

where $x(t)\in R^n$ is system state, $y(t)\in R^{n_1}$ is system output, $u(t)\in R^{n_2}$ is a known input, $w(t)\in R^{n_3}$ is an unknown input, $\sigma(t):[0,\infty)\to M=\{1,2,\ldots,m\}$ is the switching signal, which is assumed to be a piecewise continuous (from the right) function depending on time, $m\geq 2$ is the number of modes (called subsystems) of the switched system, and the system matrix quadruples (A_k,B_k,C_k,E_k), $\forall k\in M$, are assumed to be known. $k\in M$ is the active mode at time t if $\sigma(t)=k\in M$ for $t\in[t_j,t_{j+1})$ with $j=0,1,2,\ldots$, where t_j is the jth switching time instant.

Our goal is to design switched observers to estimate state x of switched system (7.1) under the assumption [31] below.

Assumption 7.1 rank $(C_k E_k) = \text{rank} E_k = n_3, \forall k \in M$.

Under Assumption 7.1, for $\forall k \in M$, we introduce the time-varying state transformation [31]:

$$T_k = \begin{bmatrix} E_k^\perp \\ (C_k E_k)^+ C_k \end{bmatrix} \triangleq \begin{bmatrix} E_k^\perp \\ \breve{T}_k \end{bmatrix}, \tag{7.2}$$

$$U_k = \begin{bmatrix} (C_k E_k)^\perp \\ (C_k E_k)^+ \end{bmatrix}. \tag{7.3}$$

Note that T_k is nonsingular and its inverse matrix is

$$T_k^{-1} = \left[(I - E_k(C_k E_k)^+ C_k)(E_k^\perp)^+, E_k\right] = \left[\tilde{T}_k, E_k\right]. \tag{7.4}$$

By using the state and output transformation

$$\bar{x} = \left[\bar{x}_1^T, \bar{x}_2^T\right]^T = T_{\sigma(t)} x, \quad \bar{y} = \left[\bar{y}_1^T, \bar{y}_2^T\right]^T = U_{\sigma(t)} y, \tag{7.5}$$

switched system (7.1) becomes in the new coordinates:

$$\begin{aligned} \dot{\bar{x}}_1 &= A_{\sigma(t),1}\bar{x}_1 + A_{\sigma(t),2}\bar{x}_2 + B_{\sigma(t),1}u, \\ \dot{\bar{x}}_2 &= A_{\sigma(t),3}\bar{x}_1 + A_{\sigma(t),4}\bar{x}_2 + B_{\sigma(t),2}u + w, \\ \bar{y}_1 &= \bar{C}_{\sigma(t)}\bar{x}_1, \\ \bar{y}_2 &= \bar{x}_2, \end{aligned} \tag{7.6}$$

where $\bar{x}_1 \in R^{n-n_3}$, $\bar{x}_2 \in R^{n_3}$, $\bar{y}_1 \in R^{n_1-n_3}$, $\bar{y}_2 \in R^{n_3}$ with

$$\begin{aligned} \bar{x}_1 &= E_{\sigma(t)}^\perp x, \\ \bar{x}_2 &= (C_{\sigma(t)} E_{\sigma(t)})^+ C_{\sigma(t)} x, \end{aligned} \tag{7.7}$$

and

$$\begin{aligned} \bar{y}_1 &= (C_{\sigma(t)} E_{\sigma(t)})^\perp y, \\ \bar{y}_2 &= (C_{\sigma(t)} E_{\sigma(t)})^+ y. \end{aligned} \tag{7.8}$$

Since transformation (7.5) depends on the switching signal, state $\bar{x}$ of the transformed switched system (7.6) may jump at switching instants t_j, which is governed by:

$$\bar{x}(t_j) = T_{\sigma(t_{j+1})} x(t_j^-) = T_{\sigma(t_{j+1})} T_{\sigma(t_j)}^{-1} \bar{x}(t_j^-), \; j = 1, \ldots. \tag{7.9}$$

From (7.6), $\bar{x}_2$ can be directly obtained through the measurement of $\bar{y}_2$. Thus, only $\bar{x}_1$ needs to be reconstructed by $\bar{y}_1$. Once state $\bar{x}_1$ is available, we can reconstruct the original state x through the inverse mapping

$$x = T_{\sigma(t)}^{-1}\bar{x} = T_{\sigma(t)}^{-1}\begin{bmatrix}\bar{x}_1\\ \bar{y}_2\end{bmatrix} = T_{\sigma(t)}^{-1}\begin{bmatrix}\bar{x}_1\\ (C_{\sigma(t)}E_{\sigma(t)})^{+}y\end{bmatrix}. \tag{7.10}$$

Now, we design switched observers only for $\bar{x}_1$.

Consider the following subsystem relevant to $\bar{x}_1$ in (7.6):

$$\dot{\bar{x}}_1 = A_{\sigma(t),1}\bar{x}_1 + A_{\sigma(t),2}\bar{x}_2 + B_{\sigma(t),1}u, \tag{7.11}$$

$$\bar{y}_1 = \bar{C}_{\sigma(t)}\bar{x}_1, \tag{7.12}$$

$$\bar{x}_1(t_j) = E_{\sigma(t_{j+1})}^{\perp}x(t_j^-),\ j = 1, 2, \ldots.$$

The proposed observer with state jumps at the switching instants has the form

$$\begin{aligned}\dot{\hat{\bar{x}}}_1 &= A_{\sigma(t),1}\hat{\bar{x}}_1 + A_{\sigma(t),2}\bar{y}_2 + B_{\sigma(t),1}u + L_{\sigma(t)}(\bar{C}_{\sigma(t)}\hat{\bar{x}}_1 - \bar{y}_1),\\ \hat{\bar{x}}_1(t_j) &= E_{\sigma(t_{j+1})}^{\perp}x(t_j^-) = E_{\sigma(t_{j+1})}^{\perp}T_{\sigma_j}^{-1}\begin{bmatrix}\hat{\bar{x}}_1(t_j^-)\\ \bar{y}_2(t_j^-)\end{bmatrix},\quad j = 1, 2, \ldots,\end{aligned} \tag{7.13}$$

where $L_k, \forall k \in M$, are the observer gains to be designed.

Define $\bar{e}_1 = \bar{x}_1 - \hat{\bar{x}}_1$ and $e = x - \hat{x}$. From (7.6), (7.12) and (7.13), the switched error system is

$$\begin{aligned}\dot{\bar{e}}_1 &= (A_{\sigma(t),1} - L_{\sigma(t)}\bar{C}_{\sigma(t)})\bar{e}_1 \triangleq \hat{A}_{\sigma(t),1}\bar{e}_1,\\ \bar{e}_1(t_j) &= E_{\sigma(t_{j+1})}^{\perp}e(t_j^-) = E_{\sigma(t_{j+1})}^{\perp}\tilde{T}_{\sigma(t_j)}\bar{e}_1(t_j^-),\ j = 1, 2, \ldots.\end{aligned} \tag{7.14}$$

Remark 7.1 In [31], the stability analysis of (7.14) was performed by a common time-invariant Lyapunov function under the strong detectability condition of (7.1), which is the invariant zeros of the matrix triplets (A_k, C_k, E_k) having negative real part. In present paper, we remove this assumption, i.e., none of the matrix pairs $(A_{k,1}, \bar{C}_k)$, $k \in M$, is required to be detectable, which is corresponding to the case that none of all subsystems of error system (7.14) is stable. Thus, our task in next section is to identify certain classes of switching signals satisfying the bound (7.18) and in the meantime to design observer gains to ensure stability of switched error system (7.14).

7.3 Main Results

In this section, to ensure stability of switched error system (7.14), first certain classes of switching signals satisfying the bound (7.18) are identified and observer gains are designed, then the switched observers for all states of system (7.1) are presented.

Theorem 7.2 *For given scalars* $\tau_{k,i}^{(1)} > 0$, $\eta_k > 0$, $\alpha_{k,l} > 0$, *and* $\varrho_{k,l} > 1$, $\forall k, i \in M$, $k \neq i$, $l = 1, 2$, *if there exist matrices* $P_{k,1} > 0$, $P_{k,2} > 0$, X_k *and* Y_k, $k \in M$, *such that the following matrix inequalities are feasible:* $\forall k, i \in M$, $k \neq i$, $l = 1, 2$, $q = 1, 2$,

$$\begin{bmatrix} -\varrho_{i,2} P_{k,1} & (E_i^{\perp} \tilde{T}_k)^T P_{i,2} \\ * & -P_{i,2} \end{bmatrix} < 0, \tag{7.15}$$

$$\begin{bmatrix} \Omega_{k,i,l}^{(q)} - \eta_k P_{k,l} & P_{k,l} - X_k - \alpha_{k,l} Y_k^T \bar{C}_k^T \\ * & -\alpha_{k,l}(X_k + X_k^T) \end{bmatrix} < 0, \tag{7.16}$$

where $\tau_{k,i}^{(2)} = \frac{\ln \varrho_{i,1}}{\eta_k}$, *and* $\Omega_{k,i,l}^{(q)} = \frac{\ln(\varrho_{k,1}\varrho_{k,2})}{\tau_{k,i}^{(1)}} P_{k,l} + P_{k,l} A_{k,l} + A_{k,l}^T P_{k,l} - Y_k \bar{C}_k - \bar{C}_k^T Y_k^T + \frac{1}{\tau_{k,i}^{(q)}}(P_{k,1} - P_{k,2})$, *then there exist a set of observer gains*

$$L_k = X_k^{-1} Y_k, k \in M, \tag{7.17}$$

such that switched observers (7.13) asymptotically estimate the states of switched system (7.11) under any switching signals with the dwell time $\tau_{k,i}$ *satisfying*

$$\tau_{k,i}^{(1)} \leq \tau_{k,i} < \tau_{k,i}^{(2)} = \frac{\ln \varrho_{i,1}}{\eta_k}, \quad \forall k, i \in M, \ k \neq i. \tag{7.18}$$

Proof The proof consists of: (i) derivation of the switched error system, (ii) computation of the time derivative of the multiple time-varying Lyapunov functions chosen as in [30] to achieve the growth restriction of Lyapunov function of the active subsystem and decrease of "energy" of the overall switched system at switching instants, and (iii) the reasoning of the convergence and stability of the overall switched error systems under the confined dwell time by a pair of upper and lower bounds.

If (7.15) and (7.16) with the unknown matrices $P_{k,1} > 0$, $P_{k,2} > 0$, X_k and Y_k, $k \in M$, are feasible, the observer gains L_k for subsystem k can be chosen as $L_k = X_k^{-1} Y_k$. Then, we obtain the closed-loop switched error system (7.14) with $L_k = X_k^{-1} Y_k$ $k \in M$.

For system (7.14), let $\ell = \{t_0, t_1, \ldots, t_j, \ldots\}$ be any sequence of switching times generated by $\sigma(t)$ satisfying (7.18). Then, we define two functions $\rho(t)$ and $\rho_1(t)$:

$$\rho(t) = \frac{t - t_j}{t_{j+1} - t_j}, \rho_1(t) = \frac{1}{t_{j+1} - t_j}, j \in N, t \in [t_j, t_{j+1}). \tag{7.19}$$

Note that $\rho_1(t)$ is a piecewise constant function on $[t_0, \infty)$. We obtain that

$$\rho(t) \in [0, 1], \rho(t_j) = 0, \rho(t_{j+1}^-) = 1. \tag{7.20}$$

For given $\varrho_{k,l} > 1, k \in M, l = 1, 2$, we then introduce a piecewise function $\varphi(t) : [t_0, \infty) \to R_{\geq 0}$:

$$\varphi(t) = (\varrho_{\sigma(t),1}\varrho_{\sigma(t),2})^{\rho(t)-1}, t \in [t_j, t_{j+1}), j \in N. \tag{7.21}$$

One can show that $\varphi(t)$ is a piecewise continuously differentiable function on $[t_0, \infty)$, and $(\varrho_{\sigma(t),1}\varrho_{\sigma(t),2})^{-1} \leq \varphi(t) \leq 1$.

To see the role of $\varphi(t)$ in the stability analysis of system (7.14), we consider the following piecewise Lyapunov function:

$$\begin{aligned} V(\bar{e}_1, t) &= \varphi(t)\bar{e}_1^T[\rho(t)P_{\sigma(t),1} + \tilde{\rho}(t)P_{\sigma(t),2}]\bar{e}_1 \\ &\overset{\Delta}{=} \varphi(t)\bar{e}_1^T P_{\sigma(t)}(t)\bar{e}_1, \end{aligned} \tag{7.22}$$

where $\tilde{\rho}(t) = 1 - \rho(t)$.

Then, one can verify that

$$\lambda_2/\varrho\|\bar{e}_1\|^2 \leq V(\bar{e}_1, t) \leq \lambda_1\|\bar{e}_1\|^2 \tag{7.23}$$

with $\varrho = \max\{\varrho_{k,1}\varrho_{k,2}, k \in M\}$, $\lambda_1 = \max\{\lambda_{\max}(P_{k,l}), k \in M, l = 1, 2\}$, and $\lambda_2 = \min\left\{\lambda_{\min}(P_{k,l}), k \in M, l = 1, 2\right\}$.

When $t \in [t_j, t_{j+1})$ and $t \in [t_{j+1}, t_{j+2})$, we set $\sigma(t) = k$ and $\sigma(t) = i$, respectively. That is, (7.14) switches from subsystem k to i at switching instant t_{j+1}. Then, when $t \in [t_j, t_{j+1})$, the time derivative of $V(\bar{e}_1, t)$ is

$$\begin{aligned} &\dot{V}(\bar{e}_1, t) \\ &= \dot{\varphi}(t)\bar{e}_1^T P_k(t)\bar{e}_1 + 2\varphi(t)\bar{e}_1^T P_k(t)\hat{A}_{k,1}\bar{e}_1 + \varphi(t)\bar{e}_1^T \rho_1(t)(P_{k,1} - P_{k,2})\bar{e}_1 \\ &\leq \varphi(t)\bar{e}_1^T \left[\frac{\ln\widehat{\mu}_k}{\tau_{k,i}^{(1)}} P_k(t) + 2P_k(t)\hat{A}_{k,1} + \frac{P_{k,1} - P_{k,2}}{t_{j+1} - t_j}\right]\bar{e}_1, \end{aligned} \tag{7.24}$$

where $\widehat{\mu}_k = \varrho_{k,1}\varrho_{k,2}$.

By pre- and post-multiplying both sides of (7.16) with $\left[I, -(L_k\bar{C}_k)^T\right]$ and $\left[I, -(L_k\bar{C}_k)^T\right]^T$, we obtain that

$$\begin{aligned} &\Omega_{k,i,l}^{(q)} - \eta_k P_{k,l} - (L_k\bar{C}_k)^T\left[P_{k,l} - X_k - \alpha_{k,l}(Y_k\bar{C}_k)^T\right]^T - P_{k,l}(L_k\bar{C}_k) \\ &+X_k(L_k\bar{C}_k) + \alpha_{k,l}\left(Y_k\bar{C}_k\right)^T(L_k\bar{C}_k) - \alpha_{k,l}(L_k\bar{C}_k)^T\left(X_k + X_k^T\right)(L_k\bar{C}_k) \\ &< 0. \end{aligned} \tag{7.25}$$

With the help of $Y_k = X_k L_k$, $\hat{A}_{k,1} = A_{k,1} - L_k\bar{C}_k$ and $\Omega_{k,l,q}$ in (7.16) with $l, q = 1, 2$, one has

$$\begin{aligned}\Theta_{k,i,l}^{(q)} &\triangleq \frac{\ln\widehat{\mu}_k}{\tau_{k,i}^{(1)}}P_{k,l} + \frac{1}{\tau_{k,i}^{(q)}}(P_{k,1} - P_{k,2}) + \hat{A}_{k,1}^T P_{k,l} + P_{k,l}\hat{A}_{k,1} \\ &< \eta_k P_{k,l}.\end{aligned} \tag{7.26}$$

Let $\rho_2(j) = \left[\frac{1}{\tau_{k,i}^{(1)}} - \frac{1}{t_{j+1}-t_j}\right] / \left[\frac{1}{\tau_{k,i}^{(1)}} - \frac{1}{\tau_{k,i}^{(2)}}\right]$ and $\tilde{\rho}_2(j) = 1 - \rho_2(j)$, $j \in N$. Since $\tau_{k,i}^{(1)} \le t_{j+1} - t_j < \tau_{k,i}^{(2)}$, we have $\rho_2(j) \in [0, 1]$. Then, one has

$$\rho_1(t) = \frac{1}{\tau_{k,i}^{(1)}}\tilde{\rho}_2(j) + \frac{1}{\tau_{k,i}^{(2)}}\rho_2(j). \tag{7.27}$$

From (7.24) together with (7.22) and (7.27), we obtain that

$$\begin{aligned}
&\dot{V}(\bar{e}_1, t) \\
&\le \varphi(t)\bar{e}_1^T\left[\frac{\ln\widehat{\mu}_k}{\tau_{k,i}^{(1)}}P_k(t) + 2P_k(t)\hat{A}_{k,1}\right]\bar{e}_1 \\
&\quad+\varphi(t)\bar{e}_1^T\left[\frac{\tilde{\rho}_2(j)}{\tau_{k,i}^{(1)}} + \frac{\rho_2(j)}{\tau_{k,i}^{(2)}}\right](P_{k,1} - P_{k,2})\bar{e}_1 \\
&= \varphi(t)\bar{e}_1^T\frac{\ln\widehat{\mu}_k}{\tau_{k,i}^{(1)}}\left(\rho(t)P_{k,1} + \tilde{\rho}(t)P_{k,2}\right)\bar{e}_1 \\
&\quad+\varphi(t)\bar{e}_1^T\left[2\left(\rho(t)P_{k,1} + \tilde{\rho}(t)P_{k,2}\right)\hat{A}_{k,1}\right]\bar{e}_1 \\
&\quad+\varphi(t)\bar{e}_1^T\left[\left(\frac{\tilde{\rho}_2(j)}{\tau_{k,i}^{(1)}} + \frac{\rho_2(j)}{\tau_{k,i}^{(2)}}\right)(P_{k,1} - P_{k,2})\right]\bar{e}_1 \\
&= \varphi(t)\bar{e}_1^T\left\{\rho(t)[\frac{\ln\hat{\mu}_k}{\tau_{k,i}^{(1)}}P_{k,1} + \hat{A}_{k,1}^T P_{k,1} + P_{k,1}\hat{A}_{k,1}]\right\}\bar{e}_1 \\
&\quad+\varphi(t)\bar{e}_1^T\left\{\tilde{\rho}(t)[\frac{\ln\hat{\mu}_k}{\tau_{k,i}^{(1)}}P_{k,2} + \hat{A}_{k,1}^T P_{k,2} + P_{k,2}\hat{A}_{k,1}]\right\}\bar{e}_1 \\
&\quad+\varphi(t)\bar{e}_1^T\left\{\frac{\tilde{\rho}_2(j)}{\tau_{k,i}^{(1)}}(P_{k,1} - P_{k,2}) + \frac{\rho_2(j)}{\tau_{k,i}^{(2)}}(P_{k,1} - P_{k,2})\right\}\bar{e}_1. \\
&= \varphi(t)\bar{e}_1^T\left(\rho(t)\hbar_{k,i}\right)\bar{e}_1 + \varphi(t)\bar{e}_1^T\left(\tilde{\rho}(t)\wp_{k,i}\right)\bar{e}_1 + \varphi(t)\bar{e}_1^T\aleph_{k,i}\bar{e}_1.
\end{aligned}$$

where $\hbar_{k,i} = \frac{\ln\hat{\mu}_k}{\tau_{k,i}^{(1)}}P_{k,1} + \hat{A}_{k,1}^T P_{k,1} + P_{k,1}\hat{A}_{k,1}$, $\wp_{k,i} = \frac{\ln\hat{\mu}_k}{\tau_{k,i}^{(1)}}P_{k,2} + \hat{A}_{k,1}^T P_{k,2} + P_{k,2}\hat{A}_{k,1}$, and $\aleph_{k,i} = \frac{\tilde{\rho}_2(j)}{\tau_{k,i}^{(1)}}(P_{k,1} - P_{k,2}) + \frac{\rho_2(j)}{\tau_{k,i}^{(2)}}(P_{k,1} - P_{k,2})$.

Since $\rho(t) + \tilde{\rho} = 1$ and $\rho_2(j) + \tilde{\rho}_2(j) = 1$, one has $\aleph_{k,i} = \rho(t)\aleph_{k,i} + \tilde{\rho}(t)\aleph_{k,i}$, $\hbar_{k,i} = \rho_2(j)\hbar_{k,i} + \tilde{\rho}_2(j)\hbar_{k,i}$, and $\wp_{k,i} = \rho_2(j)\wp_{k,i} + \tilde{\rho}_2(j)\wp_{k,i}$. Then, we obtain that

$$
\begin{aligned}
&\dot{V}(\bar{e}_1, t) \\
&\leq \phi(t)\bar{e}_1^T \left[\rho(t)\left(\rho_2(j)\hbar_{k,i} + \tilde{\rho}_2(j)\hbar_{k,i}\right)\right]\bar{e}_1 \\
&\quad +\phi(t)\bar{e}_1^T \left[\tilde{\rho}(t)\left(\rho_2(j)\wp_{k,i} + \tilde{\rho}_2(j)\wp_{k,i}\right)\right]\bar{e}_1 \\
&\quad +\phi(t)\bar{e}_1^T \left(\rho(t)\aleph_{k,i} + \tilde{\rho}(t)\aleph_{k,i}\right)\bar{e}_1 \\
&= \phi(t)\bar{e}_1^T \left[\rho(t)\left(\rho_2(j)\hbar_{k,i} + \tilde{\rho}_2(j)\hbar_{k,i} + \aleph_{k,i}\right)\right]\bar{e}_1 \\
&\quad +\phi(t)\bar{e}_1^T \left[\tilde{\rho}(t)\left(\rho_2(j)\wp_{k,i} + \tilde{\rho}_2(j)\wp_{k,i} + \aleph_{k,i}\right)\right]\bar{e}_1 \\
&= \varphi(t)\bar{e}_1^T \left[\rho(t)(\rho_2(j)\Theta_{k,i,1}^{(2)} + \tilde{\rho}_2(j)\Theta_{k,i,1}^{(1)})\right]\bar{e}_1 \\
&\quad +\varphi(t)\bar{e}_1^T \left[\tilde{\rho}(t)(\rho_2(j)\Theta_{k,i,2}^{(2)} + \tilde{\rho}_2(j)\Theta_{k,i,2}^{(1)})\right]\bar{e}_1 \\
&\leq \eta_k\varphi(t)\bar{e}_1^T \left[\rho(t)(\rho_2(j)P_{k,1} + \tilde{\rho}_2(j)P_{k,1})\right]\bar{e}_1 \\
&\quad +\eta_k\varphi(t)\bar{e}_1^T \left[\tilde{\rho}(t)(\rho_2(j)P_{k,2} + \tilde{\rho}_2(j)P_{k,2})\right]\bar{e}_1 \\
&= \eta_k\varphi(t)\bar{e}_1^T \left[\rho(t)P_{k,1} + \tilde{\rho}(t)P_{k,2}\right]\bar{e}_1 \\
&= \eta_k V(\bar{e}_1, t).
\end{aligned}
\tag{7.28}
$$

Note that the second "=" is obtained by $\Theta_{k,i,l}^{(q)}$ in (7.26), and the second "$\leq$" is obtained by (7.26). Thus, one has $V(\bar{e}_1(t), t) \leq e^{\eta_k(t-t_j)}V(\bar{e}_1(t_j), t_j)$, $t \in [t_j, t_{j+1})$, which restricts the growth of Lyapunov function of the active subsystem.

In addition, (7.15) can be written in the following equivalent form by Schur complement:

$$
(E_i^\perp \tilde{T}_k)^T P_{i,2} E_i^\perp \tilde{T}_k < \varrho_{i,2} P_{k,1}, \forall k, i \in M. \tag{7.29}
$$

At switching instant t_{j+1}, we have $\varphi(t_{j+1}) = \left(\varrho_{i,1}\varrho_{i,2}\right)^{-1}$ and $\varphi(t_{j+1}^-) = 1$. From (7.14), (7.20) and (7.29), we have

$$
\begin{aligned}
&V(\bar{e}_1(t_{j+1}), t_{j+1}) \\
&= \varphi(t_{j+1})\bar{e}_1^T(t_{j+1})\left[\rho(t_{j+1})P_{i,1} + \tilde{\rho}(t_{j+1})P_{i,2}\right]\bar{e}_1(t_{j+1}) \\
&= (\varrho_{i,1}\varrho_{i,2})^{-1}\bar{e}_1^T(t_{j+1}^-)(E_i^\perp \tilde{T}_k)^T P_{i,2} E_i^\perp \tilde{T}_k \bar{e}_1(t_{j+1}^-) \\
&< \frac{1}{\varrho_{i,1}}\bar{e}_1^T(t_{j+1}^-)P_{k,1}\bar{e}_1(t_{j+1}^-) \\
&= \frac{\varphi(t_{j+1}^-)}{\varrho_{i,1}}\bar{e}_1^T(t_{j+1}^-)\left[\rho(t_{j+1}^-)P_{k,1} + \tilde{\rho}(t_{j+1}^-)P_{k,2}\right]\bar{e}_1(t_{j+1}^-) \\
&= \frac{1}{\varrho_{i,1}}V(\bar{e}_1(t_{j+1}^-), t_{j+1}^-).
\end{aligned}
\tag{7.30}
$$

From (7.30), we know that "energy" of the overall switched system is forced to decrease at *switching instants*.

Assume that (7.14) switches from subsystem k to i at switching instant t_{j+1}. Since $\sigma(t)$ is continuous from the right and by (7.30), we have

$$
\begin{aligned}
& V(\bar{e}_1(t_{j+1}), t_{j+1}) \\
& \leq \frac{1}{\varrho_{i,1}} e^{\eta_k (t_{j+1}-t_j)} V(\bar{e}_1(t_j), t_j) \\
& = e^{-\ln \varrho_{i,1} + \eta_k (t_{j+1}-t_j)} V(\bar{e}_1(t_j), t_j).
\end{aligned} \tag{7.31}
$$

By (7.18), we conclude that $\frac{1}{\varrho_{i,1}} e^{\eta_k (t_{j+1}-t_j)} < 1$. Thus, we have that

$$
V(\bar{e}_1(t_{j+1}), t_{j+1}) < V(\bar{e}_1(t_j), t_j). \tag{7.32}
$$

Define $\eta_{max} = \max \{\eta_k, k = 1, 2, \ldots, m\}$, $\alpha_1(\|\bar{e}_1\|) = \lambda_2/\varrho \|\bar{e}_1\|^2$ and $\alpha_2(\|\bar{e}_1\|) = \lambda_1 \|\bar{e}_1\|^2$. By (7.18), there exists a finite constant τ_{max}, such that

$$
\tau_{max} \geq \frac{1}{\eta_k} \ln \varrho_{i,1}, \forall k, i \in M, k \neq i.
$$

Then, $\forall \varepsilon > 0$, we can choose

$$
\|\bar{e}_1(t_0)\| < \delta = \alpha_2^{-1}(e^{-\eta_{max}\tau_{max}} \alpha_1(\varepsilon)).
$$

Thus, it yields

$$
V(\bar{e}_1(t_0), t_0) \leq \alpha_2(\|\bar{e}_1(t_0)\|) < e^{-\eta_{max}\tau_{max}} \alpha_1(\varepsilon).
$$

Since $V(\bar{e}_1(t_j), t_j)$ is strictly decreasing, we have

$$
V(\bar{e}_1(t_j), t_j) \leq e^{-\eta_{max}\tau_{max}} \alpha_1(\varepsilon).
$$

Then,

$$
V(\bar{e}_1, t) \leq e^{\eta_i (t-t_j)} V(\bar{e}_1(t_j), t_j) \leq e^{\eta_i (t-t_j) - \eta_{max}\tau_{max}} \alpha_1(\varepsilon) \leq \alpha_1(\varepsilon),
$$

$t \in [t_j, t_{j+1})$. Hereby, it results in

$$
V(\bar{e}_1, t) < \alpha_1(\varepsilon).
$$

Furthermore, from (7.23), we can conclude

$$
\|\bar{e}_1(t)\| < \varepsilon.
$$

That is, $\forall \varepsilon > 0$, $\exists\, \delta > 0$, such that

$$
\|\bar{e}_1(t)\| < \varepsilon.
$$

Noting that the sequence $V(\bar{e}_1(t_j), t_j)$, $j = 0, 1, 2, \ldots$ is strictly decreasing, then

$$\lim_{t\to\infty} \|\bar{e}_1(t)\| = 0.$$

Therefore, switched error system (7.14) is globally asymptotically stable under switching laws $\sigma(t)$ satisfying (7.18).

Remark 7.3 The key idea behind the selection of output injection gains is: first to select the output injection gain for those individual subsystems (i.e., $L_k = X_k^{-1}Y_k$, $k \in M$), and then to identify certain classes of switching signals satisfying the bound (7.18) to make the estimation error converge to zero.

Remark 7.4 For those given parameters in the statement of Theorem 7.2, we can choose them in a "trial and error" design manner but subject to satisfaction of all the assumptions made. For L_k, we first solve the matrix inequalities (7.15) and (7.16) to obtain X_k and Y_k, then $L_k = X_k^{-1}Y_k$.

To this end, the switched observers for all states of system (7.1) are presented below.

Theorem 7.5 *Consider switched system (7.1) satisfying Assumption 7.1. For given scalars $\tau_{k,i}^{(1)} > 0$, $\eta_k > 0$, $\alpha_{k,l} > 0$, and $\varrho_{k,l} > 1$, $k, i \in M$, $i \neq k$ $l = 1, 2$, if there exist matrices $P_{k,1} > 0$, $P_{k,2} > 0$, X_k and Y_k, $k \in M$, such that (7.15)–(7.18) hold, then the switched observer*

$$\hat{x}(t) = T_{\sigma(t)}^{-1}\begin{bmatrix} \hat{\bar{x}}_1(t) \\ \bar{y}_2(t) \end{bmatrix} \tag{7.33}$$

globally asymptotically estimates the state of system (7.1) provided that dwell time $\tau_{k,i}$ satisfying (7.18), where $\bar{y}_2(t)$ is defined in (7.5), and $\hat{\bar{x}}_1(t)$ is the state of switched observer system (7.13) with $L_k = X_k^{-1}Y_k$, $k \in M$.

Proof Define $e = x - \hat{x}$. Subtracting (7.33) from (7.10) yields

$$e(t) = T_{\sigma(t)}^{-1}\begin{bmatrix} e_1(t) \\ 0 \end{bmatrix},$$

where $e_1(t)$ is the state of switched error system (7.14). From Theorem 7.2, $e_1(t)$ globally asymptotically converges to zero. Thus, $e(t)$ globally asymptotically converges to zero.

Remark 7.6 Our method does not require strong detectability of each subsystem, which nontrivially generalizes the result of [31]. From a practical point of view, our method can be promising in those applications such as the phytoplankton culture with light influence [34] with switching settings where not all subsystems are strongly detectable.

Remark 7.7 If the number of switchings is zero, the proposed switched observers cannot asymptotically estimate the state of system (7.1) since none of subsystems is strongly detectable. However, the dwell time in (7.18) is confined by a pair of upper and lower bounds, which means the activation time of each subsystem is required to be neither too long nor too short. Therefore, such switching signals can effectively avoid non-switching between subsystems, and also rule out Zeno behavior. Hence, the number of switching cannot be zero or finite when time tends to infinite ($t \to \infty$).

Remark 7.8 Note that, in terms of asymptotically estimating unknown states, the proposed switching strategy in [22] is state-dependent and specifically designed for the considered class of switched linear systems while our approach is based on the dwell time technique, which means that our switching strategy can be arbitrarily designed as long as the dwell time condition is satisfied. Note also that [22] presents a new *full-order* linear switched filter whose state variable is used to determine optimal switching control rules for switched linear systems in both continuous and discrete-time domains, and that it needs to determine $m \times (n \times n + n \times n_1)$ elements for filter matrices, (i.e., $\hat{A}_k \in R^{n\times n}$ and $\hat{B}_k \in R^{n\times n_1}$, $k \in M$) to implement the filter. In our paper, we propose a *reduced-order* observer, to implement which we only need to determine $m \times (n - n_3) \times (n_1 - n_3)$ elements for gain matrices $L_k \in R^{(n-n_3)\times(n_1-n_3)}$, $k \in M$.

In particular, we have found that if all subsystems of switched system (7.1) have the strong detectability property, then the condition on the upper bound of the dwell-time (i.e, $\tau_{k,i}^{(2)}$ in Theorem 7.2) can be removed. Thus, we have the following theorem.

Theorem 7.9 *Consider switched system (7.1) satisfying Assumption 7.1. For given scalars $\tau_{k,i} > 0$, $\eta_k > 0$, $\alpha_{k,l} > 0$, and $\varrho_{k,l} > 1$, $k, i \in M$, $l = 1, 2$, if there exist matrices $P_{k,1} > 0$, $P_{k,2} > 0$, X_k and Y_k, $k \in M$, such that the following matrix inequalities are feasible: $\forall k, i \in M$, $k \neq i$, $l = 1, 2$, $q = 1, 2$,*

$$\begin{bmatrix} -\varrho_{i,2} P_{k,1} & (E_i^{\perp}\tilde{T}_k)^T P_{i,2} \\ * & -P_{i,2} \end{bmatrix} < 0, \tag{7.34}$$

$$\begin{bmatrix} \Omega_{kil} + \eta_k P_{k,l} & P_{k,l} - X_k - \alpha_{k,l} Y_k^T \bar{C}_k^T \\ * & -\alpha_{k,l}(X_k + X_k^T) \end{bmatrix} < 0, \tag{7.35}$$

$$P_{k,1} - P_{k,2} \geq 0, \tag{7.36}$$

where $\Omega_{kil} = \frac{\ln(\varrho_{k,1}\varrho_{k,2})}{\tau_{k,i}} P_{k,l} + P_{k,l} A_{k,l} + A_{k,l}^T P_{k,l} - Y_k \bar{C}_k - \bar{C}_k^T Y_k^T + \frac{1}{\tau_{k,i}}(P_{k,1} - P_{k,2})$, *then the switched observers (7.33)*

$$\hat{x}(t) = T_{\sigma(t)}^{-1} \begin{bmatrix} \hat{\bar{x}}_1(t) \\ \bar{y}_2(t) \end{bmatrix} \tag{7.37}$$

globally asymptotically estimates the state of system (7.1) with the mode-dependent dwell time $\tau_{k,i}$, *where* $\bar{y}_2(t)$ *is defined in (7.5), and* $\hat{\bar{x}}_1(t)$ *is the state of switched observer system (7.13) with* $L_k = X_k^{-1} Y_k$, $k \in M$.

Proof The proof can be easily derived by the methodology as above. Therefore, the proof of Theorem 7.9 is omitted in here.

Remark 7.10 When $P_{k,1} = P_{k,2} = P, \forall k \in M$, in Theorem 7.9 (that is, a common Lyapunov function is used), the conditions of [31] are recovered in the following two cases, if the parameters are chosen as $\kappa_2 = \eta_k > 0$, $\alpha = \alpha_{k,l} > 0$, $k \in M$, $l = 1, 2$.

Case 1: $\varrho_1 = \varrho_{k,l} > 1$, and $\tau_1 = \tau_{k,i} > 0$, $k, i \in M$, $k \neq i$, $l = 1, 2$. From (7.35), we can obtain that

$$\frac{\ln \varrho_1^2}{\tau_1} P + \left(A_{k,1} - L_k \bar{C}_k\right)^T P + P\left(A_{k,1} - L_k \bar{C}_k\right) < -\kappa_2 P,$$

which means that

$$\dot{V}(\bar{e}_1, t) < -\kappa_2 V(\bar{e}_1, t).$$

Denote $\kappa_1 = 1/\varrho_1 < 1$. Then $\ln \kappa_1 < 0$ and $\tau_1 > \frac{\ln \kappa_1}{\kappa_2}$. By (7.34), one can prove that

$$V(\bar{e}_1(t_{j+1}), t_{j+1}) \leq \kappa_1 V(\bar{e}_1(t_{j+1}^-), t_{j+1}^-),$$

which means that $V(\bar{e}_1(t), t)$ is strictly decreasing at the switching times. Thus, one concludes that $V(\bar{e}_1, t)$ converges exponentially to zero as t tends to infinity, which implies that $\bar{e}_1(t)$ converges exponentially to zero, if the dwell time τ satisfies that $\tau \geq \tau_1 > \frac{\ln \kappa_1}{\kappa_2}$.

Case 2: $\varrho_1 = \varrho_{k,1}$, $\varrho_2 = \varrho_{k,2}$, $0 < \varrho_1 < 1$, and $\varrho_1 \varrho_2 > 1$. Obviously, $\tau_1 = \tau_{k,i} = \frac{\ln \kappa_1}{\kappa_2} > 0$, $\forall k, i \in M$, with $\kappa_1 = \frac{1}{\varrho_1}$. We also have

$$\dot{V}(\bar{e}_1, t) < -\kappa_2 V(\bar{e}_1, t).$$

By (7.34), one has that

$$V(\bar{e}_1(t_{j+1}), t_{j+1}) \leq \kappa_1 V(\bar{e}_1(t_{j+1}^-), t_{j+1}^-).$$

By Theorem 7.9, $\bar{e}_1(t)$ converges exponentially to zero if the dwell time $\tau \geq \tau_1 = \frac{\ln \kappa_1}{\kappa_2}$.

Remark 7.11 In terms of using a common Lyapunov function P as stated in Remark 7, the difference between [31] and Theorem 7.9 is: in [31] P is first computed and then the dwell time is obtained, while in our paper a dwell time is pre-given and then check if there exists a feasible solution P to the matrix inequalities for such a dwell time.

7.4 Application to Stabilization of Switched Singular Linear Systems

Switched singular systems have also drawn considerable attention in the recent past [35–43]. As stated in [44], it is necessary to allow for solutions in switched singular systems with instantaneous state jumps, which are unavoidable even if all subsystems are regular and impulse-free. This is one of the major distinctions between switched singular systems and switched conventional systems [43]. Thus, in turn, considerable research attention has been devoted to the crucial property of stability under arbitrary switchings and under some constrained switching laws [36, 38–45]. It should be noted that the dwell-time-based methods in [36, 38–43] require that there must be at least one stable subsystem of the switched system to be switched on for the stability analysis to be successful. Thus, naturally questions arise: Is it possible, in the framework of the dwell time technique, to achieve the stabilization design of switched singular linear systems *without posing any stability requirements on subsystems of the switched systems*? If possible, then under which conditions and how can we come up with a switching policy to achieve this goal? To the best of our knowledge, in the literature there are no results which provide answers to these questions.

In this subsection, we will apply Theorem 7.2 to provide answers to these questions. First, a time-varying coordinate transformation is introduced first in order to convert the problem into an equivalent one of for reduced-order switched conventional linear system with state jumps. Then, by constructing certain new multiple time-varying Lyapunov functions, computable sufficient conditions for the global stabilization task are proposed within the framework of a dwell-time switching. Given the assumed instability of individual subsystems, the stabilization of the switched system is achieved under the condition of confining the dwell time by a certain pair of upper and lower bounds, which restrict the growth of Lyapunov function for the actively operating subsystem, thus decrease energy of the overall switched system at switching times. In addition, the multiple time-varying Lyapunov functions method is also used to analyze the stability analysis of a class of switched linear singular systems with stable subsystems.

7.4.1 System Description

Consider the following switched linear singular systems:

$$\begin{cases} E_{\sigma(t)}\dot{x} = A_{\sigma(t)}x, \\ x(0) = x_0, \end{cases} \tag{7.38}$$

where $x \in R^n$ is the system state, $x_0 \in R^n$ is a vector-valued initial state, $\sigma(t)$: $R^+ \to M = \{1, 2, \ldots, m\}$ is the switching law, which is assumed to be a piecewise continuous (from the right) function of time and $m > 0$ is the number of modes of the switched system (i.e., subsystems). Throughout this paper, we assume that $\sigma(t) = \sigma(t_k) = i_k, i_k \in M, t \in [t_k, t_{k+1})$, where t_k is the switching instant, this means that the $i_k th$ subsystem is activated when $t \in [t_k, t_{k+1})$. For every $i \in M$, E_i and A_i are constant matrices, and it is assumed that $rank(E_i) = r \le n$. For simplicity, we use (E_i, A_i) to denote the ith subsystem. The set $\{t_k\}$ generated by $\sigma(t) \in T[\tau_1, \tau_2]$ denotes the switching sequences with $\tau_1 \le t_k - t_{k-1} \le \tau_2, k \in N$.

Definition 7.12 [46]. For every $i \in M$, the singular system (E_i, A_i) is said to be

(i) regular if $\det(sE_i - A_i)$ is not identically zero;
(ii) impulse-free if $\deg(\det(sE_i - A_i)) = rank(E_i)$.

Assumption 7.2 For every $i \in M$, the singular system (E_i, A_i) is regular and impulse-free.

Due to the fact $rank(E_i) = r \le n$, we can find nonsingular matrices H_i and N_i $(i \in M)$, such that

$$H_i E_i N_i = \begin{bmatrix} I_r & 0 \\ 0 & 0 \end{bmatrix} =: \bar{E}, \tag{7.39}$$

$$H_i A_i N_i = \begin{bmatrix} A_{11}(i) & A_{12}(i) \\ A_{21}(i) & A_{22}(i) \end{bmatrix} =: \bar{A}_i. \tag{7.40}$$

By introducing the state transformation:

$$\bar{x} = \begin{bmatrix} \bar{x}_1 \\ \bar{x}_2 \end{bmatrix} = N_i^{-1} x, \ t \in [t_k, t_{k+1}), \tag{7.41}$$

switched system (7.38) takes the following form in the new coordinates:

$$\begin{cases} \dot{\bar{x}}_1 = A_{11}(\sigma(t))\bar{x}_1 + A_{12}(\sigma(t))\bar{x}_2, \\ 0 = A_{21}(\sigma(t))\bar{x}_1 + A_{22}(\sigma(t))\bar{x}_2. \end{cases} \tag{7.42}$$

Note that at the switching instant t_k, the system switches from (E_j, A_j) to (E_i, A_i). Then, considering the switching law dependent feature of the state transformation (7.41), we have

$$x(t_k^-) = N_j \bar{x}(t_k^-), x(t_k^+) = N_i \bar{x}(t_k^+). \tag{7.43}$$

According to the analysis presented in [43], we have that

$$\bar{x}(t_k^+) = \Gamma_{ij} \bar{x}(t_k^-), i, j \in M, \tag{7.44}$$

with

$$\Gamma_{ij} = \begin{bmatrix} I & 0 \\ -A_{22}^{-1}(i)A_{21}(i) & 0 \end{bmatrix} N_i^{-1} N_j. \tag{7.45}$$

In addition, we can obtain that

$$\begin{aligned}\bar{x}_1\left(t_k^+\right) &= \left[I\ 0\right]\Gamma_{ij}\bar{x}\left(t_k^-\right)\\ &= \left[I\ 0\right]\Gamma_{ij}\begin{bmatrix} I \\ -A_{22}^{-1}(j)A_{21}(j)\end{bmatrix}\bar{x}_1\left(t_k^-\right)\\ &= \Pi_{ij}\bar{x}_1\left(t_k^-\right),\end{aligned} \tag{7.46}$$

with

$$\Pi_{ij} = \left[I\ 0\right]N_i^{-1}N_j\begin{bmatrix} I \\ -A_{22}^{-1}(j)A_{21}(j)\end{bmatrix}. \tag{7.47}$$

Under Assumption 7.2, we know that A_{22} is nonsingular. Thus, by (7.42), we can obtain a reduced-order switched conventional linear system with state jumps (7.46):

$$\begin{cases}\dot{\bar{x}}_1 = \hat{A}\left(\sigma(t)\right)\bar{x}_1,\\ \bar{x}_1\left(t_k^+\right) = \Pi_{ij}\bar{x}_1\left(t_k^-\right),\end{cases} \tag{7.48}$$

where $\hat{A}\left(\sigma(t)\right) = A_{11}\left(\sigma(t)\right) - A_{12}\left(\sigma(t)\right)A_{22}^{-1}\left(\sigma(t)\right)A_{21}(\sigma(t))$.

Therefore, to derive sufficient conditions for the existence of the switching signal $\sigma(t)$ to globally asymptotically stabilize the switched singular linear system (7.38) is converted into the globally asymptotically stabilization of the switched linear system (7.48) by designing switching signal $\sigma(t)$. However, it is easy to see that the stabilization design for system (7.48) is equivalent to the one for system (7.14). Now, we apply Theorem 7.2 to solve the stabilization design for system (7.48). We consider two classes of switched singular systems. First we consider the stabilization problem for a switched singular system with all unstable subsystems. And thereafter we study the stability analysis problem for a switched singular system with all stable subsystems for the sake of establishing a parallel.

7.4.2 Case A: All Subsystems Are Unstable

Theorem 7.13 *Consider switched singular system (7.38) satisfying Assumption 7.2. If there exist constants $\tau_2 \geq \tau_1 > 0$, $\lambda > 0$, $1 > \mu > 0$, $\mu_i > 1$, $i \in M$, and positive definite matrices $P_{i1} > 0$, $P_{i2} > 0$, and any appropriate dimensional matrix P_{i3} and P_{i4}, such that*

$$\theta_{ilq} - \lambda P_{il} \leq 0,\ \forall i \in M,\ l, q = 1, 2, \tag{7.49}$$

$$\begin{bmatrix} -\mu P_{j1} & \Pi_{ij}^T P_{i2} \\ * & -\mu_i P_{i2}\end{bmatrix} < 0, \forall i, j \in M, i \neq j, \tag{7.50}$$

$$\ln\mu + \lambda\tau_2 < 0, \tag{7.51}$$

where

$$\theta_{ilq} = \vartheta_i P_{il} + \hat{A}^T(i) P_{il} + P_{il}\hat{A}(i) + \frac{P_{i1} - P_{i2}}{\tau_q}, \forall i \in M,\ l, q = 1, 2, \quad (7.52)$$

$$\Pi_{ij} = \begin{bmatrix} I & 0 \end{bmatrix} N_i^{-1} N_j \begin{bmatrix} I \\ -A_{22}^{-1}(j) A_{21}(j) \end{bmatrix} \quad (7.53)$$

with

$$\vartheta_i = \frac{\ln(\mu_i)}{\tau_1}, \ \hat{A}(i) = A_{11}(i) - A_{12}(i) A_{22}^{-1}(i) A_{21}(i),$$

then switched singular system (7.38) is globally uniformly asymptotically stable under switching law $\sigma(t) \in T[\tau_1, \tau_2]$.

Proof For $\{t_k\}$ generated by $\sigma(t) \in T[\tau_1, \tau_2]$ and $t \in [t_k, t_{k+1})$, we define:

$$\rho(t) = \frac{t - t_k}{t_{k+1} - t_k}, \tilde{\rho}(t) = 1 - \rho(t), \rho_1(t) = \frac{1}{t_{k+1} - t_k}, \phi(t) = \mu_i^{\rho(t)-1}. \quad (7.54)$$

When $t \in [t_k, t_{k+1})$, we consider the following Lyapunov function:

$$V_i(t, x) = \phi(t) \bar{x}^T \bar{E} \bar{P}_i(t) \bar{x}, \quad (7.55)$$

where $\bar{P}_i(t) = \begin{bmatrix} P_i(t) & 0 \\ P_{i3} & P_{i4} \end{bmatrix}$ with $P_i(t) = \rho(t) P_{i1} + \tilde{\rho}(t) P_{i2}, i \in M$.

From (7.55), we can obtain that

$$\begin{aligned} & V_i(t, x) \\ &= \phi(t) \bar{x}^T \bar{E} \bar{P}_i(t) \bar{x} \\ &= \phi(t) \begin{bmatrix} \bar{x}_1^T & \bar{x}_2^T \end{bmatrix} \begin{bmatrix} I_r & 0 \\ 0 & 0 \end{bmatrix} \begin{bmatrix} P_i(t) & 0 \\ P_{i3} & P_{i4} \end{bmatrix} \begin{bmatrix} \bar{x}_1 \\ \bar{x}_2 \end{bmatrix} \\ &= \phi(t) \begin{bmatrix} \bar{x}_1^T & 0 \end{bmatrix} \begin{bmatrix} P_i(t) & 0 \\ P_{i3} & P_{i4} \end{bmatrix} \begin{bmatrix} \bar{x}_1 \\ \bar{x}_2 \end{bmatrix} \\ &= \phi(t) \begin{bmatrix} \bar{x}_1^T P_i(t) & 0 \end{bmatrix} \begin{bmatrix} \bar{x}_1 \\ \bar{x}_2 \end{bmatrix} \\ &= \phi(t) \bar{x}_1^T P_i(t) \bar{x}_1 \\ &= V_i(t, \bar{x}_1). \end{aligned} \quad (7.56)$$

It is obvious that $V_i(t, \bar{x}_1)$ satisfies

$$\alpha_1(\|\bar{x}_1\|) = \frac{\lambda_2}{\nu} \|\bar{x}_1\|^2 \le V_i(t, \bar{x}_1) \le \lambda_1 \|\bar{x}_1\|^2 = \alpha_2(\|\bar{x}_1\|), \quad (7.57)$$

where $\nu = \max\{\mu_i, i \in M\}$, $\lambda_1 = \max\{\lambda_{\max}(P_{il}), i \in M, l = 1, 2\}$, and $\lambda_2 = \min\{\lambda_{\min}(P_{il}), i \in M, l = 1, 2\}$.

When $t \in [t_k, t_{k+1})$, the time derivative of $V_i(t, \bar{x}_1)$ is

$$\begin{aligned}
&\dot{V}_i(t, x_1) \\
&= \dot{\rho}(t) \ln \mu_i \phi(t) \bar{x}_1^T P_i(t) \bar{x}_1 + \phi(t) \bar{x}_1^T \left[\dot{\rho}(t) P_{i1} + \dot{\tilde{\rho}}(t) P_{i2}\right] \bar{x}_1 + 2\phi(t) \bar{x}_1^T P_i(t) \dot{\bar{x}}_1 \\
&\leq \phi(t) \bar{x}_1^T \left[\frac{\ln \mu_i}{\tau_1} P_i(t) + \rho_1(t)(P_{i1} - P_{i2})\right] \bar{x}_1 + 2\phi(t) \bar{x}_1^T P_i(t) \hat{A}(i) \bar{x}_1 \\
&= \phi(t) \bar{x}_1^T \left[\frac{\ln \mu_i}{\tau_1} P_i(t) + P_i(t) \hat{A}(i) + \hat{A}^T(i) P_i(t) + \rho_1(t)(P_{i1} - P_{i2})\right] \bar{x}_1. \quad (7.58)
\end{aligned}$$

We choose a function $\rho_2(t) \in [0, 1]$ and $\rho_2(t) = 1 - \tilde{\rho}_2(t)$, such that

$$\rho_1(t) = \frac{1}{\tau_1} \rho_2(t) + \frac{1}{\tau_2} \tilde{\rho}_2(t). \quad (7.59)$$

Apparently, such a function $\rho_2(t)$ can be fairly easy obtained. For example, when $\tau_2 > \tau_1$, we choose $\rho_2(t) = \left(\rho_1(t) - \frac{1}{\tau_2}\right) / \left(\frac{1}{\tau_1} - \frac{1}{\tau_2}\right)$. In addition, if $\tau_1 = \tau_2$, then $\rho_1(t) = \tau_1 = \tau_2$, we can easily to choose, for example, $\rho_2(t) = \tilde{\rho}_2(t) = \frac{1}{2}$, which satisfies (7.59).

From (7.58) and (7.59), one has that

$$\begin{aligned}
&\dot{V}_i(t, \bar{x}_1) \\
&\leq \phi(t) \bar{x}_1^T \left[\vartheta_i P_i(t) + P_i(t) \hat{A}(i) + \hat{A}^T(i) P_i(t) + \rho_1(t)(P_{i1} - P_{i2})\right] \bar{x}_1 \\
&= \phi(t) \bar{x}_1^T \left\{\rho(t) [\vartheta_i P_{i1} + \hat{A}^T(i) P_{i1} + P_{i1} \hat{A}(i)]\right\} \bar{x}_1 \\
&\quad + \phi(t) \bar{x}_1^T \left\{\tilde{\rho}(t) [\vartheta_i P_{i2} + \hat{A}^T(i) P_{i2} + P_{i2} \hat{A}(i)]\right\} \bar{x}_1 \\
&\quad + \phi(t) \bar{x}_1^T \left[(\frac{1}{\tau_1} \rho_2(t) + \frac{1}{\tau_2} \tilde{\rho}_2(t))(P_{i1} - P_{i2})\right] \bar{x}_1. \quad (7.60)
\end{aligned}$$

With the help of (7.49) and (7.60), we can obtain

$$\begin{aligned}
&\dot{V}_i(t, \bar{x}_1) \\
&\leq \phi(t) \bar{x}_1^T \left\{\rho(t) \left[\rho_2(t) \theta_{i11} + \tilde{\rho}_2(t) \theta_{i12}\right]\right\} \bar{x}_1 \\
&\quad + \phi(t) \bar{x}_1^T \left\{\tilde{\rho}(t) \left[\rho_2(t) \theta_{i21} + \tilde{\rho}_2(t) \theta_{i22}\right]\right\} \bar{x}_1 \\
&\leq \phi(t) \bar{x}_1^T \left\{\rho(t) \left[\rho_2(t) \lambda P_{i1} + \tilde{\rho}_2(t) \lambda P_{i1}\right]\right\} \bar{x}_1 \\
&\quad + \phi(t) \bar{x}_1^T \left\{\tilde{\rho}(t) \left[\rho_2(t) \lambda P_{i2} + \tilde{\rho}_2(t) \lambda P_{i2}\right]\right\} \bar{x}_1 \\
&= \lambda \phi(t) \bar{x}_1^T \left[\rho(t) P_{i1} + \tilde{\rho}(t) P_{i2}\right] \bar{x}_1 \\
&= \lambda V_i(t, \bar{x}_1). \quad (7.61)
\end{aligned}$$

On the other hand, according to (7.50), one can find that

$$\begin{aligned}
& V_i\left(t_k^+, \bar{x}_1\right) \\
&= \phi\left(t_k^+\right) \bar{x}_1^T\left(t_k^+\right) P_i\left(t_k^+\right) \bar{x}_1\left(t_k^+\right) \\
&= \mu_i^{\rho\left(t_k^+\right)-1} \bar{x}_1^T\left(t_k^+\right)\left[\rho\left(t_k^+\right) P_{i1} + \tilde{\rho}\left(t_k^+\right) P_{i2}\right] \bar{x}_1\left(t_k^+\right) \\
&= \mu_i^{-1} \bar{x}_1^T\left(t_k^+\right) P_{i2} \bar{x}_1\left(t_k^+\right) \\
&= \mu_i^{-1} \bar{x}_1^T\left(t_k^-\right) \Pi_{ij}^T P_{i2} \Pi_{ij} \bar{x}_1\left(t_k^-\right) \\
&< \bar{x}_1^T\left(t_k^-\right) \mu P_{j1} \bar{x}_1\left(t_k^-\right) \\
&= \mu \mu_i^{\rho\left(t_k^-\right)-1} \bar{x}_1^T\left(t_k^-\right) P_{j1} \bar{x}_1\left(t_k^-\right) \\
&= \mu \bar{x}_1^T\left(t_k^-\right)\left[\rho\left(t_k^-\right) P_{j1} + \tilde{\rho}\left(t_k^-\right) P_{j2}\right] \bar{x}_1\left(t_k^-\right) \\
&= \mu \phi\left(t_k^-\right) \bar{x}_1^T\left(t_k^-\right) P_j\left(t_k^-\right) \bar{x}_1\left(t_k^-\right) \\
&= \mu V_j\left(t_k^-, \bar{x}_1\right).
\end{aligned} \tag{7.62}$$

For simplicity, we define

$$V\left(t, \bar{x}_1\right) = \sum_{i=1}^{m} \alpha_i(t) V_i\left(t, \bar{x}_1\right),$$

where $\alpha_i(t) = \begin{cases} 1, & if \sigma(t) = i, \\ 0, & otherwise, \end{cases}$ $\forall i \in M$. Assuming $\alpha_i(t) = 1$ and $\alpha_j(t) = 0$, when $t \in [t_k, t_{k+1})$.

From (7.61), we can derive $V_i\left(t, \bar{x}_1\right) \leq e^{\lambda(t-t_k)} V_i\left(t_k, \bar{x}_1\right), t \in [t_k, t_{k+1})$.

Then, by supposing system (7.48) switches from subsystem j to i at switching instant t_k, since $\sigma(t)$ is continuous from the right, we have

$$V_i\left(t_k, \bar{x}_1\right) \leq \mu e^{\lambda(t_k - t_{k-1})} V_j\left(t_{k-1}, \bar{x}_1\right). \tag{7.63}$$

Since $t_k - t_{k-1} \leq \tau_2$, $k = 1, 2, \ldots$, which together with (7.51), we can obtain that

$$\mu e^{\lambda(t_k - t_{k-1})} < 1. \tag{7.64}$$

Thus, we can see $V_i\left(t_k, \bar{x}_1\right) < V_j\left(t_{k-1}, \bar{x}_1\right)$. Then, for any $\varepsilon > 0$, we can choose $\left\|\bar{x}_1\left(t_0\right)\right\| < \delta\left(\varepsilon\right) = \alpha_2^{-1}\left(e^{-\lambda\tau_2}\alpha_1\left(\varepsilon\right)\right)$. Thus, this yields $V\left(t_0, \bar{x}_1\right) \leq \alpha_2\left(\left\|\bar{x}_1\left(t_0\right)\right\|\right) < e^{-\lambda\tau_2}\alpha_1\left(\varepsilon\right)$. Since $V_i\left(t_k, \bar{x}_1\right)$ is strictly decreasing, we have $V_i\left(t_k, \bar{x}_1\right) \leq e^{-\lambda\tau_2}\alpha_1\left(\varepsilon\right)$. Then, we have $V\left(t, \bar{x}_1\right) \leq \alpha_1\left(\varepsilon\right)$. Furthermore, from (7.57), we can conclude $\left\|\bar{x}_1(t)\right\| < \varepsilon$. Obviously, for $\forall \delta > 0$, we have $\left\|\bar{x}_1(t)\right\| < \varepsilon$. Due to the fact that the sequence $V_i\left(t_k, \bar{x}_1\right), k = 0, 1, 2, \ldots$ is strictly decreasing, we obtained that $\lim_{t \to \infty} \left\|\bar{x}_1(t)\right\| = 0$. Therefore, we can conclude switched system (7.48) is globally asymptotically stabilized under switching law $\sigma(t) \in T[\tau_1, \tau_2]$.

It follows from (7.42) that $\bar{x}_2(t) = -A_{22}^{-1}(\sigma(t))\, A_{21}(\sigma(t))\, \bar{x}_1(t)$, thus, $\bar{x}_2(t)$ also global asymptotical stable. This indicates that system (7.42), or equivalently, the system (7.38) is globally asymptotically stabilize under switching law $\sigma(t) \in T[\tau_1, \tau_2]$. Thus, the proof is completed.

Remark 7.14 Theorem 7.13 provides a sufficient condition to achieve the stabilization for a class of switched singular systems via dwell-time-based switchings. Our method does not require stability of each subsystem, which nontrivially generalizes the result of [43] obtained under the assumption that all or part of subsystems are stable. On the other hand, when $E_i = I, \forall i \in M$, the switched singular system deduces to the switched normal system in [47]. Therefore, our obtained result extends that of the switched normal system to the switched singular system case.

Remark 7.15 In our results, the common λ and μ are used to confine accordingly the upper bound of the dwell time τ_2. Since these two parameters may be dependent on individual subsystems, values of these two parameters can be different for different subsystems. It is therefore that here proposed methods appears less conservative and thus yield a better outcome design result.

7.4.3 *Case B: All Subsystems Are Stable*

It is well known that switching among stable systems may yield instability of the overall switched system. It is therefore that is necessary also to investigate this line of design derivation for the case where all subsystems of switched singular systems are stable. In particular, we have found that if the subsystems of switched singular linear system (7.38) are stable, then the condition on the upper bound of the dwell-time (i.e,τ_2 in Theorem 7.13) can be removed. Thus, we have the following theorem.

Theorem 7.16 *Consider switched singular system (7.38) satisfying Assumption 7.2. If there exist constants* $\tau_1 > 0, \lambda > 0, 1 > \mu > 0, \mu_i > 1, i \in M$, *and positive definite matrices* $P_{i1} > 0,\ P_{i2} > 0$, *such that*

$$\theta_{ilq} + \lambda P_{il} \le 0,\ \forall i \in M,\ l, q = 1, 2, \tag{7.65}$$

$$\begin{bmatrix} -\mu P_{j1} & \Pi_{ij}^T P_{i2} \\ * & -\mu_i P_{i2} \end{bmatrix} < 0, \forall i, j \in M, i \neq j, \tag{7.66}$$

$$P_{i1} - P_{i2} > 0, \tag{7.67}$$

where

$$\theta_{ilq} = \vartheta_i P_{il} + \hat{A}^T(i) P_{il} + P_{il}\hat{A}(i) + \frac{1}{\tau_1}(P_{i1} - P_{i2}), \forall i \in M,\ l, q = 1, 2, \tag{7.68}$$

$$\Pi_{ij} = \begin{bmatrix} I & 0 \end{bmatrix} N_i^{-1} N_j \begin{bmatrix} I \\ -A_{22}^{-1}(j)A_{21}(j) \end{bmatrix}, \tag{7.69}$$

with $\vartheta_i = \frac{\ln \mu_i}{\tau_1}$, *and* $\hat{A}(i) = A_{11}(i) - A_{12}(i)A_{22}^{-1}(i)A_{21}(i)$, *then switched singular system (7.38) is globally uniformly asymptotically stable under switching law* $\sigma(t) \in T[\tau_1, \infty)$.

Proof The proof is very similar to the proof of Theorem 7.13, and it can be easily derived by the methodology as above. Therefore, the proof of Theorem 7.16 is omitted in here.

7.5 Illustrative Examples

7.5.1 Example 1

In this subsection, an example is studied to show the effectiveness of switched observers design for switched linear systems (7.1) with two subsystems:

$$A_1 = \begin{bmatrix} -1 & 0 & -1 & 1 & 1 \\ 1 & -1 & 1 & -2 & 1 \\ 1 & 1 & -1 & -1 & -0.1 \\ 1 & -1 & 0 & -1 & 1 \\ 1 & 1 & 3 & 1 & -1 \end{bmatrix}, E_1^T = \begin{bmatrix} 1 & 1 \\ 0 & 1 \\ 0 & 1 \\ 0 & 0 \\ 0 & 1 \end{bmatrix},$$

$$A_2 = \begin{bmatrix} 1 & 0 & 0 & 1 & 1 \\ 1 & -1 & 1 & 2 & 1 \\ 1 & 1 & -1 & 1 & 1 \\ 1 & 0 & -1 & -1 & 0 \\ 1 & -1 & 0 & -1 & -1 \end{bmatrix}, E_2^T = \begin{bmatrix} 0 & 0 \\ 0 & 1 \\ 1 & 1 \\ 0 & 0 \\ 0 & 1 \end{bmatrix},$$

$$B_1 = B_2 = 0,$$

$$C_1 = \begin{bmatrix} -1 & 0 & 0 & 0 & 0 \\ 0 & 0 & 0 & 0 & 0 \\ 0 & 1 & 0 & 0 & 0 \end{bmatrix}, C_2 = \begin{bmatrix} 0 & 0 & 0 & 1 & 0 \\ 0 & 1 & 1 & 0 & 0 \\ -1 & 0 & 1 & 1 & 0 \end{bmatrix}.$$

Set $w(t) = [2\sin(5t) + 3, 4\cos(6t) - 5]^T$. Since $\operatorname{rank} E_k = \operatorname{rank}(C_k E_k) = 2$, $k = 1, 2$, Assumption 7.2 holds. Following the design procedure in Sect. 7.3, we obtain the state transformation matrices:

$$T_1 = \begin{bmatrix} 0 & -1 & 1 & 0 & 0 \\ 0 & 0 & 0 & 1 & 0 \\ 0 & -1 & 0 & 0 & 1 \\ 1 & -1 & 0 & 0 & 0 \\ 0 & 1 & 0 & 0 & 0 \end{bmatrix}, \quad T_2 = \begin{bmatrix} 1 & 0 & 0 & 0 & 0 \\ 0 & 0 & 0 & 1 & 0 \\ 0 & -1 & 0 & 0 & 1 \\ -2 & -1 & 1 & 2 & 0 \\ 1 & 1 & 0 & -1 & 0 \end{bmatrix},$$

$$U_1 = \begin{bmatrix} 0 & 1 & 0 \\ -1 & 0 & -1 \\ 0 & 0 & 1 \end{bmatrix}, U_2 = \begin{bmatrix} 1 & 0 & 0 \\ 0 & -1 & 2 \\ 0 & 1 & -1 \end{bmatrix}.$$

Then, in the new coordinates switched system (7.6) is obtained with the following:

$$\left[\begin{array}{c:c} A_{1,1} & A_{1,2} \\ \hdashline A_{1,3} & A_{1,4} \end{array}\right] = \left[\begin{array}{ccc:cc} -2 & 1 & -1.1 & 0 & -2.1 \\ 0 & -1 & 1 & 1 & 1 \\ 2 & 3 & -2 & 0 & 2 \\ \hdashline -2 & 3 & 0 & -2 & -3 \\ 1 & -2 & 1 & 1 & 2 \end{array}\right],$$

$$\left[\begin{array}{c:c} A_{2,1} & A_{2,2} \\ \hdashline A_{2,3} & A_{2,4} \end{array}\right] = \left[\begin{array}{ccc:cc} 0 & 2 & 1 & 0 & 1 \\ 0 & 0 & 0 & -1 & -1 \\ 1 & -4 & -2 & -1 & -3 \\ \hdashline -4 & -1 & -2 & -4 & -4 \\ 2 & 3 & 2 & 2 & 3 \end{array}\right],$$

$$\bar{C}_1 = \begin{bmatrix} 0 & 0 & 0 \end{bmatrix}, \ \bar{C}_2 = \begin{bmatrix} 0 & 1 & 0 \end{bmatrix}.$$

We can also obtain switched error system (7.14) with

$$E_2^{\perp}\tilde{T}_1 = \begin{bmatrix} 0 & 0 & 0 \\ 0 & 1 & 0 \\ 0 & 0 & 1 \end{bmatrix}, E_1^{\perp}\tilde{T}_2 = \begin{bmatrix} 2 & 0 & 0 \\ -2 & 1 & 0 \\ 0 & 0 & 1 \end{bmatrix}.$$

Based on the observability decomposition (which can be done by 'obsvf' in MATLAB) and the definition of strong detectability, one can obtain that the simple detectability of both the matrix pairs $(A_{1,1}, \bar{C}_1)$ and $(A_{2,1}, \bar{C}_2)$ is satisfied, and then the strong detectability of the studied system is not satisfied. Therefore, the methods in [29–32] cannot be applied to design switched observers for the studied switched system. Because the proposed method in present paper does not require detectability of subsystems of switched linear systems (7.1), we can apply it to this example to asymptotically estimate the state of the switched system, as shown below.

Let $\eta_1 = \eta_2 = 10$, $\varrho_{1,1} = \varrho_{2,1} = 1.39$, $\varrho_{1,2} = \varrho_{2,2} = 1.01$, $\alpha_{k,l} = 0.05$, $k = 1, 2$, $l = 1, 2$, and $\tau_{1,2}^{(1)} = \tau_{2,1}^{(1)} = 0.02\ s$. By Theorem 7.16, we obtain observer gains for each subsystem:

$$L_1 = \begin{bmatrix} 0, 0, 0 \end{bmatrix}^T,$$

$$L_2 = \begin{bmatrix} 3.0443, 46.0056, -18.2445 \end{bmatrix}^T.$$

It should be noted that each subsystem of switched error system is *unstable*, since the eigenvalues of $A_{1,1} - L_1\bar{C}_1$ and $A_{2,1} - L_2\bar{C}_2$ are $\{0.2162, -2.6081 \pm 1.2351i\}$ and $\{0.4142, -2.4142, -46.0056\}$, respectively.

By Theorem 7.16, the proposed switched observer can globally asymptotically estimate the state of system (7.1) under any switching law $\sigma(t)$ with $0.02s \leq \tau_{1,2} < \frac{\ln 1.39}{10}s$ and $0.02s \leq \tau_{2,1} < \frac{\ln 1.39}{10}s$.

Let $e(t_0) = (1.6, -2.8, -3.4)^T$. From Fig. 7.1, we see that both subsystems of switched error system are unstable. Figure 7.2a shows the state trajectories of the whole dynamics of the observation error under a randomly chosen switching signal $\sigma(t)$ with $0.02s \leq \tau_{1,2} < \frac{\ln 1.39}{10}s$ and $0.02s \leq \tau_{2,1} < \frac{\ln 1.39}{10}s$. The switching signal is shown in Fig. 7.2b. It can be clearly observed from Fig. 7.2a that asymptotical stability of the switched error system has been achieved. Thus, the simulation results well show the effectiveness of the proposed method.

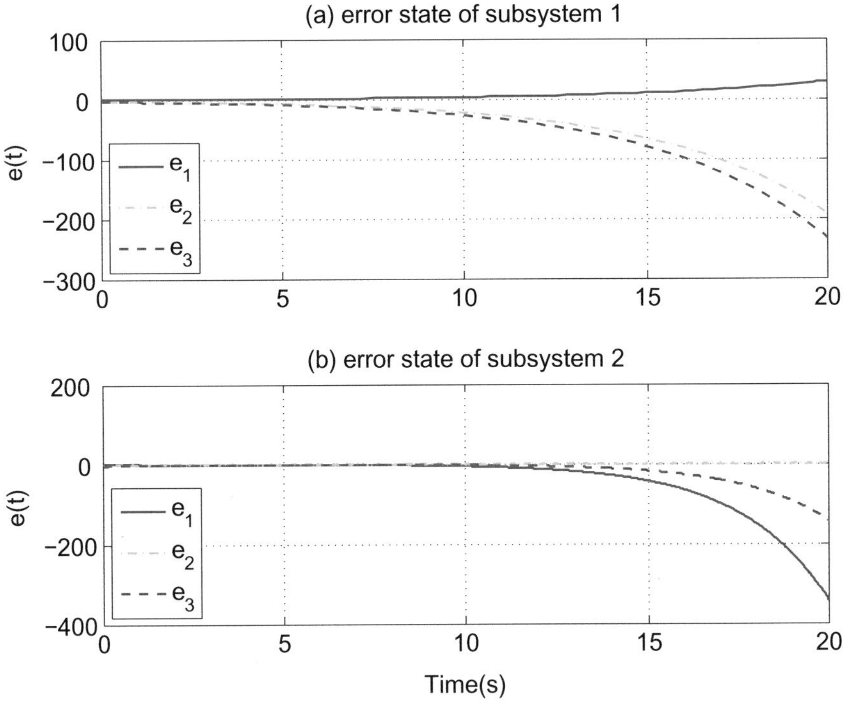

Fig. 7.1 The trajectories of the error state e of subsystems

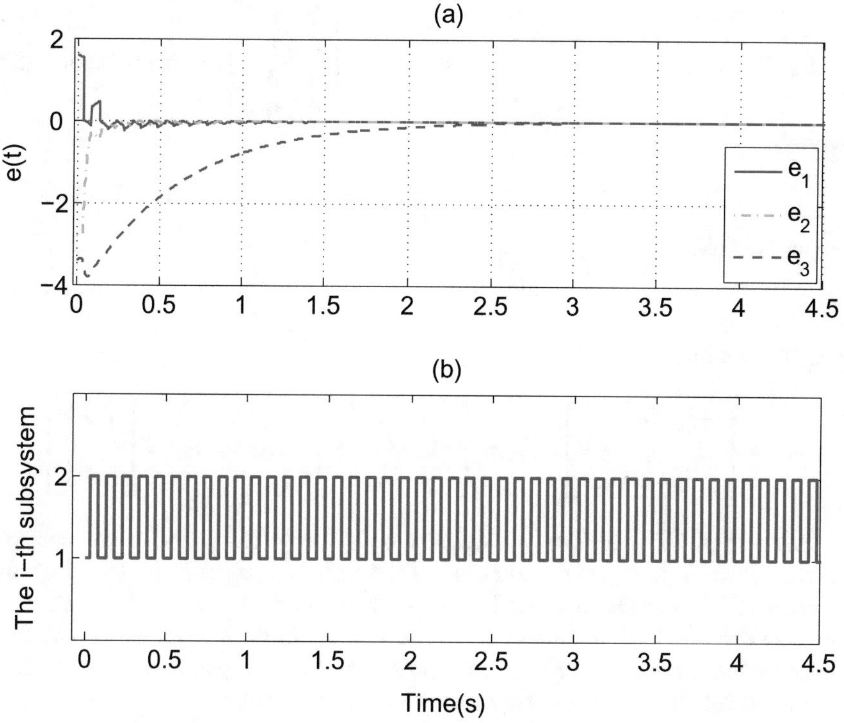

Fig. 7.2 Trajectories and switching signal of the switched error system

7.5.2 Example 2

Consider the following switched singular linear system (7.38) with two subsystems of the fourth order:

$$E_1 = \begin{bmatrix} 1\,0\,0\,0 \\ 0\,1\,0\,0 \\ 0\,0\,1\,0 \\ 0\,0\,0\,0 \end{bmatrix}, A_1 = \begin{bmatrix} -1 & 1 & 0 & -1 \\ 1 & -2 & -1 & 1 \\ 1 & -3 & -0.2 & -0.2 \\ 2 & 1 & -2 & -1 \end{bmatrix},$$
$$E_2 = \begin{bmatrix} 1\,0\,0\,0 \\ 0\,0\,1\,0 \\ 0\,1\,0\,0 \\ 0\,0\,0\,0 \end{bmatrix}, A_2 = \begin{bmatrix} 1 & -2 & -1 & 2 \\ -2 & -2 & -1 & 2 \\ -1 & 1 & -1 & -2 \\ 1 & 1 & -1 & 1 \end{bmatrix}.$$

By $H_1 = N_1 = N_2 = \begin{bmatrix} 1\,0\,0\,0 \\ 0\,1\,0\,0 \\ 0\,0\,1\,0 \\ 0\,0\,0\,1 \end{bmatrix}$ and $H_2 = \begin{bmatrix} 1\,0\,0\,0 \\ 0\,0\,1\,0 \\ 0\,1\,0\,0 \\ 0\,0\,0\,1 \end{bmatrix}$, we can obtain (7.39) with

$$\bar{A}_1 = H_1 A_1 N_1 = \begin{bmatrix} -1 & 1 & 0 & -1 \\ 1 & -2 & -1 & 1 \\ 1 & -3 & -0.2 & -0.2 \\ 2 & 1 & -2 & -1 \end{bmatrix}, \bar{A}_2 = H_2 A_2 N_2 = \begin{bmatrix} 1 & -2 & -1 & 2 \\ -1 & 1 & -1 & -2 \\ -2 & -2 & -1 & 2 \\ 1 & 1 & -1 & 1 \end{bmatrix},$$

and (7.48) with

$$\hat{A}_1 = \begin{bmatrix} -3 & 0 & 2 \\ 3 & -1 & -3 \\ 0.6 & -3.2 & 0.2 \end{bmatrix}, \hat{A}_2 = \begin{bmatrix} -1 & -4 & 1 \\ 1 & 3 & -3 \\ -4 & -4 & 1 \end{bmatrix}, \Pi_{12} = \Pi_{21} = \begin{bmatrix} 1\,0\,0 \\ 0\,1\,0 \\ 0\,0\,1 \end{bmatrix}.$$

From Figs. 7.3 and 7.4, we see that both subsystem 1 and 2 are unstable with $x(t_0) = (3, 4, -2, 14)^T$ and $x(t_0) = (3, 4, -2, 9)^T$, respectively. Therefore, the method in [43] cannot be applied to the studied switched system. However, since the here proposed method does not require stability of the subsystems of switched singular linear system (7.38), we can apply it to this example in order to stabilize asymptotically the switched system, as shown further below.

Let $\lambda = 1$, $\mu = 0.9$, $\mu_1 = 2.1$, $\mu_2 = 2.2$, $\tau_1 = 0.08s$, $\tau_2 = 0.1s$. Then by means of Theorem 7.13, we obtain the following matrices:

$$P_{11} = \begin{bmatrix} 57.9402 & 55.2065 & -21.1590 \\ 55.2065 & 60.0027 & -18.8876 \\ -21.1590 & -18.8876 & 28.4523 \end{bmatrix},$$

$$P_{12} = \begin{bmatrix} 56.6759 & 58.1763 & -25.8976 \\ 58.1763 & 67.8901 & -32.9528 \\ -25.8976 & -32.9528 & 38.5350 \end{bmatrix},$$

$$P_{21} = \begin{bmatrix} 44.8649 & 46.5320 & -9.4428 \\ 46.5320 & 53.1657 & -12.1667 \\ -9.4428 & -12.1667 & 22.8090 \end{bmatrix},$$

$$P_{22} = \begin{bmatrix} 83.2191 & 76.1833 & -47.7394 \\ 76.1833 & 80.6093 & -46.3264 \\ -47.7394 & -46.3264 & 52.6676 \end{bmatrix}.$$

Let $x(t_0) = (3, 4, -2, 14)^T$. Figure 7.5 shows the state trajectories of the switched system under the dwell time switching signal $\sigma(t) \in T[\tau_1, \tau_2)$ shown in Fig. 7.6. Apparently Fig. 7.5 shows that the asymptotical stability of the switched system has been achieved rather quickly. Thus, the simulation results demonstrate pretty well the effectiveness of the proposed switching design method.

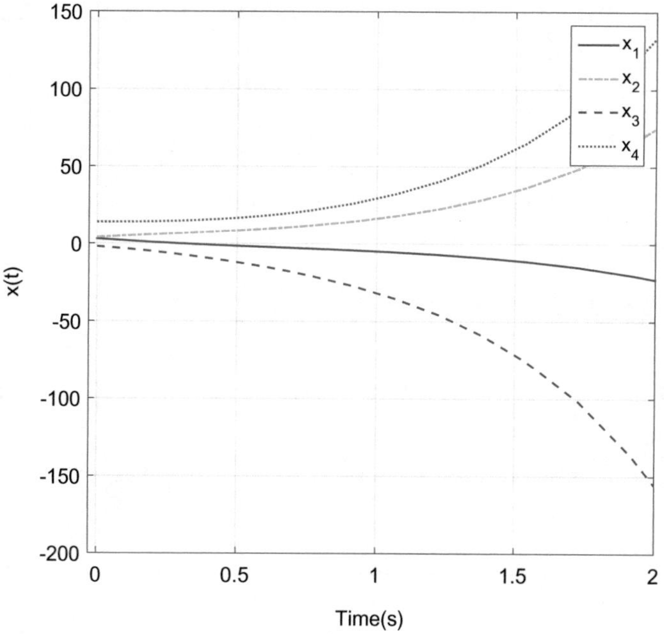

Fig. 7.3 The trajectories of subsystem 1

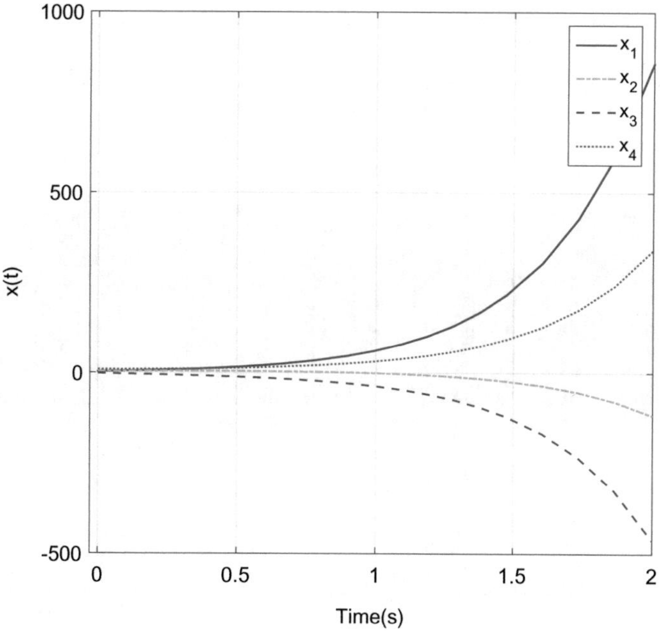

Fig. 7.4 The trajectories of subsystem 2

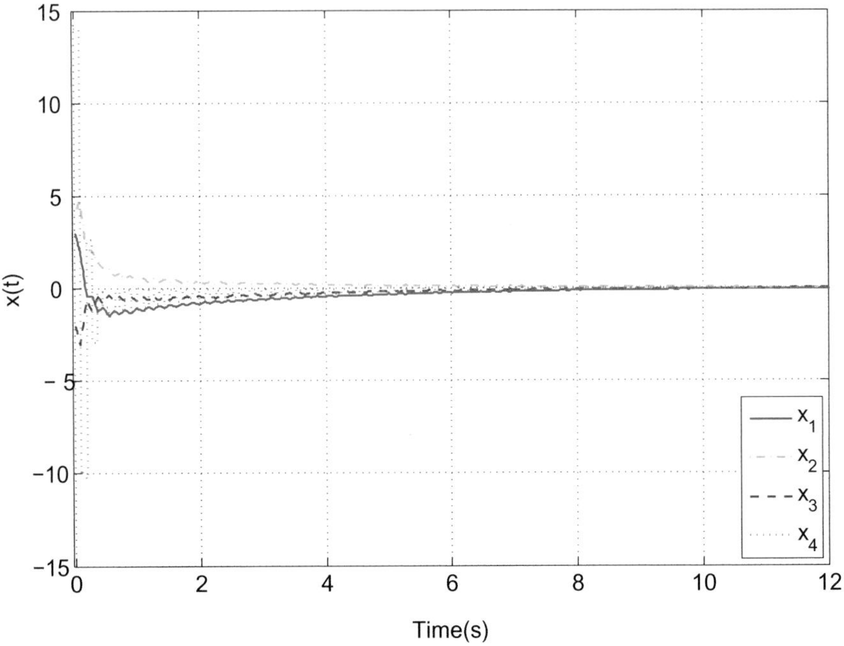

Fig. 7.5 State responses of the switched singular system

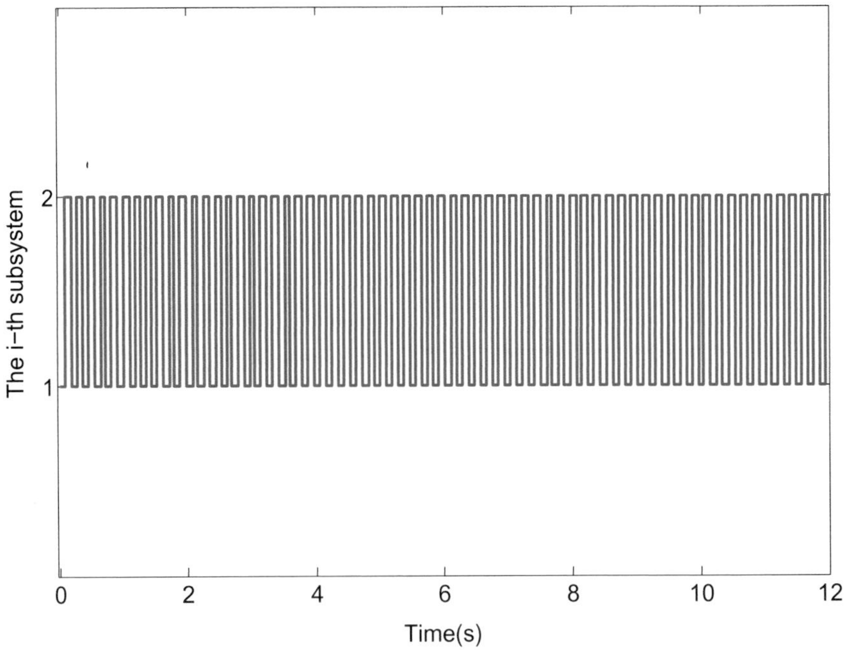

Fig. 7.6 The designed switching signal $\sigma(t)$

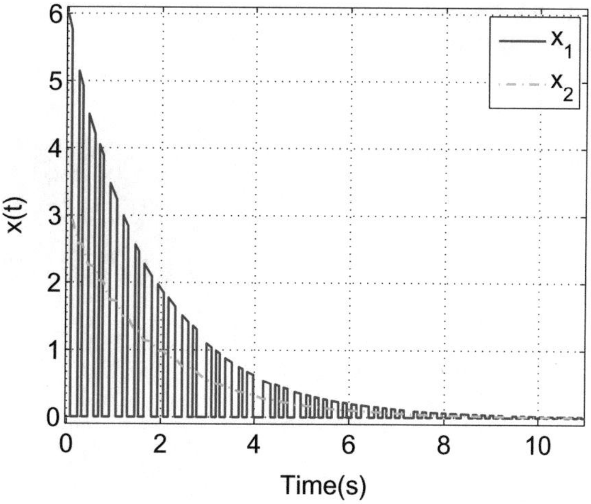

Fig. 7.7 State response of the switched singular system

7.5.3 Example 3

Consider the same switched singular system as in [43], where

$$E_1 = \begin{bmatrix} 0 & 1 \\ 0 & 0 \end{bmatrix}, A_1 = \begin{bmatrix} 0 & -1 \\ 1 & a \end{bmatrix}, E_2 = \begin{bmatrix} 1 & 1 \\ 0 & 0 \end{bmatrix}, A_2 = \begin{bmatrix} -1 & -1 \\ 1 & 0 \end{bmatrix}.$$

As in [43], we set $N_1 = \begin{bmatrix} 0 & 1 \\ 1 & 0 \end{bmatrix}$, $N_2 = \begin{bmatrix} 0 & 1 \\ 1 & -1 \end{bmatrix}$, then get $\bar{E} = E_1 N_1 = E_2 N_2 = \begin{bmatrix} 1 & 0 \\ 0 & 0 \end{bmatrix}$, $\bar{A}_1 = A_1 N_1 = \begin{bmatrix} -1 & 0 \\ a & 1 \end{bmatrix}$, $\bar{A}_2 = A_2 N_2 = \begin{bmatrix} -1 & 0 \\ 0 & 1 \end{bmatrix}$.

When $a = -2$, it is reported in [43] that when the average dwell time is greater than $\frac{1}{2} \ln (|1 - a|) = 0.5493$, the system is asymptotical stability.

Now, we apply our Theorem 7.16 to the studied system, and it has been found that, when $a = -2$, there exist $\tau_1 = 0.05s$, $\lambda = 10$, $\mu_1 = \mu_1 = 1.002$, and $P_{11} = [11.0254]$, $P_{12} = [7.0712]$, $P_{21} = [3.6752]$, $P_{22} = [2.3893]$, such that (7.65), (7.66) and (7.67) hold. Thus, the studied switched singular system is asymptotically stable under switching law $\sigma(t) \in T\,[0.05, \infty)$. When $a = -2$, the state response of the switched singular system and the corresponding switching signal are illustrated in Fig. 7.7 and Fig. 7.8, respectively.

Remark 7.17 It can be seen that the dwell time 0.05 obtained by Theorem 7.16 is much smaller than the one $\frac{1}{2} \ln (|1 - a|) = 0.5493$ in [43]. Thus, our obtained stability results are less conservative than the ones given in [43].

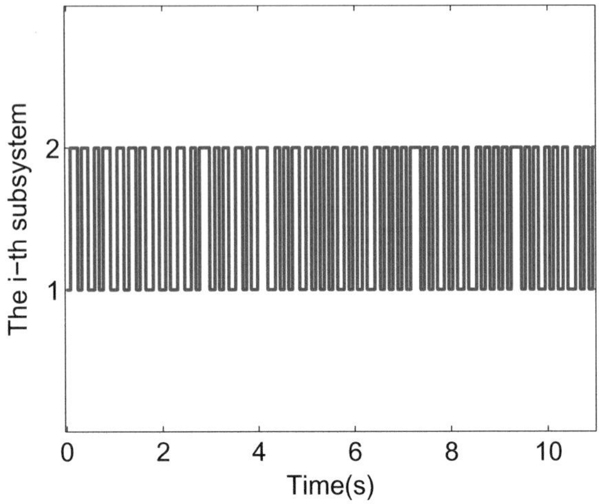

Fig. 7.8 The corresponding switching signal

7.6 Conclusion

This paper has investigated state observers design problem of a class of unknown inputs switched linear systems via mode-dependent dwell time switchings *without requiring strong detectability condition of subsystems* of the switched systems. Since the observer of individual subsystem cannot be designed due to unavailability of strong detectability condition of the subsystem, the state of the switched system is estimated under the condition of confining the dwell time by a pair of upper and lower bounds, restricting the growth of Lyapunov function of the active subsystem, and forcing "energy" of the overall switched system to decrease at switching instants. Next, we apply our theorem to the stabilization of switched singular linear systems.

Future works will focus on controller design problem with the estimated state information generated by the proposed observer in this manuscript.

References

1. Liberzon, D.: Switching in Systems and Control. Birkhauser, Boston (2003)
2. Branicky, M.S.: Multiple lyapunov functions and other analysis tools for switched and hybrid systems. IEEE Trans. Autom. Control **43**(4), 475–482 (1998)
3. Lin, H., Antsaklis, P.J.: Stability and stabilizability of switched linear systems: a survey of recent results. IEEE Trans. Autom. Control **54**(2), 308–322 (2009)
4. Sun, Z., Ge, S.S.: Switched Linear Systems: Control and Design. Springer, London (2005)
5. Shi, P., Su, X., Li, F.: Dissipativity-based filtering for fuzzy switched systems with stochastic perturbation. IEEE Trans. Autom. Control **61**(6), 1694–1699 (2016)
6. Wu, L., Zheng, W.: Weighted h_∞ model reduction for linear switched systems with time-varying delay. Automatica **45**(1), 186–193 (2009)
7. Wu, L., Yang, R., Shi, Peng, Su, X.: Stability analysis and stabilization of 2-d switched systems under arbitrary and restricted switchings. Automatica **59**, 206–215 (2015)

8. Wang, J.-N., Lanzon, A., Petersen, I.R.: Robust cooperative control of multiple heterogeneous negative-imaginary systems. Automatica **61**, 64–72 (2015)
9. Zhang, L., Gao, H.: Asynchronously switched control of switched linear systems with average dwell time. Automatica **46**(5), 953–958 (2010)
10. Zhai, G., Hu, B., Yasuda, K., Michel, A.N.: Stability analysis of switched systems with stable and unstable subsystems: an average dwell time approach. Int. J. Syst. Sci. **32**, 1055–1061 (2001)
11. Yang, H., Jiang, B., Zhao, J.: On finite-time stability of cyclic switched nonlinear systems. IEEE Trans. Autom. Control **60**(8), 2201–2206 (2015)
12. Sun, X.-M., Wang, W.: Integral input-to-state stability for hybrid delayed systems with unstable continuous dynamics. Automatica **48**, 2359–2364 (2012)
13. Fu, J., Jin, Y., Chai, T.: Fault-tolerant control of a class of switched nonlinear systems with structural uncertainties. IEEE Trans. Circuits Syst. II: Express Briefs **PP**(99), 1–1 (2015)
14. Fu, J., Ma, R., Chai, T.: Global finite-time stabilization of a class of switched nonlinear systems with the powers of positive odd rational numbers. Automatica **54**(4), 360–373 (2015)
15. Ma, R., Zhao, J.: Backstepping design for global stabilization of switched nonlinear systems in lower triangular form under arbitrary switchings. Automatica **46**(11), 1819–1823 (2010)
16. Zhao, X., Zheng, X., Niu, B., Liu, L.: Adaptive tracking control for a class of uncertain switched nonlinear systems. Automatica **52**(2), 185–191 (2015)
17. Zhao, J., Hill, D.J.: On stability, L_2-gain and H_∞ control for switched systems. Automatica **44**(5), 1220–1232 (2008)
18. Zhao, J., Hill, D.J.: Dissipativity theory for switched systems. IEEE Trans. Autom. Control **53**(4), 941–953 (2008)
19. Colaneri, P., Middleton, R.H., Chen, Z., Caporale, D., Blanchini, F.: Convexity of the cost functional in an optimal control problem for a class of positive switched systems. Automatica **50**(4), 1227–1234 (2014)
20. Chiang, M.-L., Fu, L.-C.: Adaptive stabilization of a class of uncertain switched nonlinear systems with backstepping control. Automatica **50**(8), 2128–2135 (2014)
21. Santarelli, K.R., Megretski, A., Dahleh, M.A.: Stabilizability of two-dimensional linear systems via switched output feedback. Syst. Control Lett. **57**(3), 228–235 (2008)
22. Geromel, J.C., Colaneri, P., Bolzern, P.: Dynamic output feedback control of switched linear systems. IEEE Trans. Autom. Control **53**(3), 720–733 (2008)
23. Alessandri, A., Coletta, P.: Switching observers for continuous-time and discrete-time linear systems. In: Proceedings of the 2001 American Control Conference, vol. 3, pp. 2516–2521 (2001)
24. Tanwani, A., Shim, H., Liberzon, D.: Observability for switched linear systems: characterization and observer design. IEEE Trans. Autom. Control **58**(4), 891–904 (2013)
25. Wu, J., Sun, Z.: Observer-driven switching stabilization of switched linear systems. Automatica **49**(8), 2556–2560 (2013)
26. Xie, G., Wang, L.: Periodic stabilizability of switched linear control systems. Automatica **45**(9), 2141–2148 (2009)
27. Yang, J., Chen, Y., Zhu, F., Yu, K., Bu, X.: Synchronous switching observer for nonlinear switched systems with minimum dwell time constraint. J. Frankl. Inst. **352**(11), 4665–4681 (2015)
28. Zhao, X., Liu, H., Zhang, J., Li, H.: Multiple-mode observer design for a class of switched linear systems. IEEE Trans. Autom. Sci. Eng. **12**(1), 272–280 (2015)
29. Defoort, M., Van Gorp, J., Djemai, M., Veluvolu, K.: Hybrid observer for switched linear systems with unknown inputs. In: 7th IEEE Conference on Industrial Electronics and Applications, pp. 594–599 (2012)
30. Huang, G.-J., Chen, W.-H.: A revisit to the design of switched observers for switched linear systems with unknown inputs. Int. J. Control Autom. Syst. **12**(5), 954–962 (2014)
31. Bejarano, F.J., Pisano, A.: Switched observers for switched linear systems with unknown inputs. IEEE Trans. Autom. Control **56**(3), 681–686 (2011)

32. Yang, J., Zhu, F., Tan, X., Wang, Y.: Robust full-order and reduced-order observers for a class of uncertain switched systems. J. Dyn. Syst. Meas. Control **138**(2) (2016)
33. Chen, J., Li, J., Yang, S., Deng, F.: Weighted optimization-based distributed kalman filter for nonlinear target tracking in collaborative sensor networks. IEEE Trans. Cybern. https://doi.org/10.1109/TCYB.2016.2587723
34. Moreno, J.A., Rocha-Cózatl, E., Wouwer, A.V.: A dynamical interpretation of strong observability and detectability concepts for nonlinear systems with unknown inputs: application to biochemical processes. Bioprocess Biosyst. Eng. **37**, 37–49 (2014)
35. Gao, Z., Yang, K., Shen, Y., Ji, Z.: Input-to-state stability analysis for a class of nonlinear switched descriptor systems. Int. J. Syst. Sci. **46**(16), 2973–2981 (2015)
36. Liberzon, D., Trenn, S.: Switched nonlinear differential algebraic equations: solution theory, lyapunov functions, and stability. Automatica **48**(5), 954–963 (2012)
37. Lin, J.-X., Fei, S.-M.: Robust exponential admissibility of uncertain switched singular time-delay systems. Acta Automatica Sinica **36**(12), 1773–1779 (2010)
38. Lin, J., Fei, S., Gao, Z.: Stabilization of discrete-time switched singular time-delay systems under asynchronous switching. J. Frankl. Inst. **349**(5), 1808–1827 (2012)
39. Mu, X., Wei, J., Ma, R.: Stability of linear switched differential algebraic equations with stable and unstable subsystems. Int. J. Syst. Sci. **44**(10), 1879–1884 (2013)
40. Wang, Y., Zou, Y., Liu, Y., Shi, X., Zuo, Z.: Average dwell time approach to finite-time stabilization of switched singular linear systems. J. Frankl. Inst. **352**(7), 2920–2933 (2015)
41. Zamani, I., Shafiee, M.: On the stability issues of switched singular time-delay systems with slow switching based on average dwell-time. Int. J. Robust Nonlinear Control **24**(4), 595–624 (2014)
42. Zamani, I., Shafiee, M., Ibeas, A.: Stability analysis of hybrid switched nonlinear singular time-delay systems with stable and unstable subsystems. Int. J. Syst. Sci. **45**(5), 1128–1144 (2014)
43. Zhou, L., Ho, D.W.C., Zhai, G.: Stability analysis of switched linear singular systems. Automatica **49**(5), 1481–1487 (2013)
44. Liberzon, D., Trenn, S.: On stability of linear switched differential algebraic equations. In: Proceedings of the 48th IEEE Conference on Decision and Control held jointly with the 28th Chinese Control Conference, pp. 2156–2161 (2009)
45. Raouf, J., Michalska, H.: Exponential stabilization of singular systems by controlled switching. In: Proceedings of the 49th IEEE Conference on Decision and Control, pp. 414–419 (2010)
46. Lam, J., Shengyuan, X.: Robust Control and Filtering of Singular Systems. Springer, Berlin/Heidelberg (2006)
47. Xiang, W., Xiao, J.: Stabilization of switched continuous-time systems with all modes unstable via dwell time switching. Automatica **50**(3), 940–945 (2014)

Chapter 8
Conclusions

The research of stabilization and H_∞ control for switched linear systems is an interesting and meaningful work. Many practical systems can be modeled as switched systems. The introduction of switching mechanism leads to complicated system behaviors. Even though the subsystems are linear systems, switched linear systems could exhibit nonlinear characteristics under different switching signals. Hence, it is of great theoretical and practical application significance to analyze and synthesize of switched systems. This book has presented several interesting theories for switched systems. The main contributions of this book are concluded as following:

In Chap. 2, the global finite-time stabilization problem for a class of switched nonlinear systems under arbitrary switchings is investigated. All subsystems of the studied switched system under consideration are in lower triangular form. Based on the adding one power integrator technique, both a class of non-Lipschitz continuous state feedback controller and a common Lyapunov function are simultaneously constructed such that the closed-loop switched system is global finite-time stability under arbitrary switchings. In the controller design process, a common coordinate transformation of all subsystems is exploited to avoid using individual coordinate transformation for each subsystem. Finally, Two examples are given to show the effectiveness of the proposed method.

In Chap. 3, the global finite-time stabilization is studied for a class of switched strict-feedback nonlinear systems, whose subsystems have chained integrators with the powers of positive odd rational numbers (i.e., numerators and denominators of the powers are all positive odd integers but not necessarily relatively prime). All the powers in each equation of subsystems of the switched systems can be different. Based on the technique of adding a power integrator, the global finite-time stabilizers of individual subsystems are first systematically constructed to guarantee global finite-time stability of the closed-loop smooth switched system under arbitrary switchings, and then a co-design of stabilizers and a state-dependent switching law is proposed

J. Fu and R. Ma, *Stabilization and H_∞ Control of Switched Dynamic Systems*, Studies in Systems, Decision and Control 310,
https://doi.org/10.1007/978-3-030-54197-2_8

to achieve global finite-time stabilization of the closed-loop non-smooth switched systems. In the controller design, a common coordinate transformation of all subsystems is exploited to avoid using individual coordinate transformations for individual subsystems. We also give some sufficient conditions that enable our design by characterizing the powers of the chained integrators of the considered switched systems. Numerical examples are provided to demonstrate the effectiveness of the proposed results.

In Chap. 4, the global finite-time stabilization is investigated for a class of switched nonlinear systems in non-triangular form, whose subsystems have chained integrators with the powers of positive odd rational numbers (i.e., numerators and denominators of the powers are all positive odd integers). All subsystems are not assumed to be stabilizable. Based on the technique of adding a power integrator and the multiple Lyapunov functions method, both the global finite-time stabilizers of individual subsystems and a switching law are systematically constructed to guarantee global finite-time stabilization of the closed-loop switched nonlinear system. A numerical example is provided to illustrate the effectiveness of the proposed method.

In Chap. 5, the global adaptive finite-time stabilization is investigated by logic-based switching control for a class of uncertain nonlinear systems with the powers of positive odd rational numbers. Parametric uncertainties entering the state equations nonlinearly can be fast time-varying or jumping at unknown time instants, and the control coefficient appearing in the control channel can be unknown. The bounds of the parametric uncertainties and the unknown control coefficient are not required to know a priori. Our proposed controller is a switching-type one, in which a nonlinear controller with two parameters to be tuned is first designed by adding a power integrator, and then a switching mechanism is proposed to tune the parameters online to finite-time stabilize the system. An example is provided to demonstrate the effectiveness of the proposed result.

In Chap. 6, the standard H_∞ control of switched systems is studied via dwell time switchings without posing any internal stability requirements on subsystems of the switched systems. First, a sufficient condition is formed by specifying lower and upper bounds of the dwell time, constraining upper bound of derivative of Lyapunov function of the active subsystem, and forcing the Lyapunov function values of the overall switched system to decrease at switching times to achieve standard H_∞ control of unforced switched linear systems. Then, in the same framework of the dwell time, sufficient conditions are given for that of the corresponding forced switched linear systems by further designing state feedback controllers. Finally, numerical examples are provided to demonstrate the effectiveness of the proposed results.

In Chap. 7, the state observers design of a class of unknown inputs switched linear systems is investigated via mode-dependent dwell time switchings. The distinguishing feature of the proposed method is that strong detectability condition of subsystems of the switched systems is unnecessarily required. Firstly, a time-varying coordinate transformation is introduced to design a suitable reduced-order observer for each subsystem. Then, computable sufficient conditions on the synthesis of the observers are proposed in the framework of a mode-dependent dwell time technique. Since the observer of individual subsystem cannot be designed due to unavailability of strong detectability condition of the subsystem, the state of the switched system is estimated under the condition of confining the dwell time by a pair of upper and lower bounds, restricting the growth of Lyapunov function of the active subsystem, and forcing "energy" of the overall switched system to decrease at *switching instants*. Next, we apply our method to the stabilization of switched singular linear systems. Finally, examples are presented to demonstrate the effectiveness of the proposed methods.

Printed in the United States
by Baker & Taylor Publisher Services